DATE DUE

Advances in

ECOLOGICAL RESEARCH

VOLUME 39

Advances in Ecological Research

Series Editor: HAL CASWELL
Biology Department
Woods Hole Oceanographic Institution
Woods Hole, Massachusetts

This publication has been supported by La Trobe University
Internet: http://www.latrobe.edu.au

Advances in

ECOLOGICAL RESEARCH

VOLUME 39

Floods in an Arid Continent

Edited by
ALDO POIANI

Faculty of Science, Technology and Engineering
La Trobe University, Mildura
Victoria, Australia

2006

AMSTERDAM • BOSTON • HEIDELBERG • LONDON
NEW YORK • OXFORD • PARIS • SAN DIEGO
SAN FRANCISCO • SINGAPORE • SYDNEY • TOKYO
Academic Press is an imprint of Elsevier

ELSEVIER

Academic Press is an imprint of Elsevier
525 B Street, Suite 1900, San Diego, California 92101-4495, USA
84 Theobald's Road, London WC1X 8RR, UK

This book is printed on acid-free paper. ∞

For information on all Academic Press publications
visit our Web site at www.books.elsevier.com

ISBN-13: 978-0-12-373630-7
ISBN-10: 0-12-373630-7

PRINTED IN THE UNITED STATES OF AMERICA
06 07 08 09 9 8 7 6 5 4 3 2 1

Contributors to Volume 39

DEBBIE ABBS, *CSIRO Marine and Atmospheric Research, Aspendale, Victoria 3195, Australia.*

CATHERINE ALLAN, *School of Environmental and Information Sciences, Charles Sturt University, Albury, New South Wales 2640, Australia.*

JOSEPH AZUOLAS, *Department of Primary Industries, Attwood, Victoria 3049, Australia.*

ANDREA BALLINGER*, *Australian Centre for Biodiversity: Analysis, Policy and Management, School of Biological Sciences, Monash University, Clayton, Victoria 3800, Australia.*

NICK BOND, *School of Biological Sciences, Monash University, Clayton, Victoria 3800, Australia.*

ALLAN CURTIS, *Institute for Land, Water and Society, Faculty of Science and Agriculture, Charles Sturt University, Albury, New South Wales 2640, Australia.*

JOHN FREEBAIRN, *Melbourne Institute of Applied Economics and Social Research, The University of Melbourne, Parkville, Victoria 3010, Australia.*

ROGER JONES, *CSIRO Marine and Atmospheric Research, Aspendale, Victoria 3195, Australia.*

SAM LAKE, *School of Biological Sciences, Monash University, Clayton, Victoria 3800, Australia.*

RALPH MAC NALLY, *Australian Centre for Biodiversity: Analysis, Policy and Management, School of Biological Sciences, Monash University, Clayton, Victoria 3800, Australia.*

NICKI MAZUR, *ENVision Environmental Consulting, Hackett, Australian Capital Territory 2602, Australia; Institute for Land, Water and Society, Charles Sturt University, Albury, New South Wales 2640, Australia.*

JOHN MORTON, *Sociology and Anthropology, La Trobe University, Bundoora, Victoria 3086, Australia.*

BARRIE PITTOCK, *CSIRO Marine and Atmospheric Research, Aspendale, Victoria 3195, Australia.*

ALDO POIANI, *Faculty of Science, Technology and Engineering, La Trobe University, Mildura, Victoria 3502, Australia.*

*Current address: Department of Sustainability and Environment, Melbourne, Victoria 3002, Australia.

PAUL REICH, *School of Biological Sciences, Monash University, Clayton, Victoria 3800, Australia.*

RAMASAMY SUPPIAH, *CSIRO Marine and Atmospheric Research, Aspendale, Victoria 3195, Australia.*

ANDREW J. TURNER, *Andrew Turner Consulting Pty Ltd., Glen Lee, Princes Hill, Victoria 3054, Australia.*

COLIN R. WILKS, *School of Veterinary Science, The University of Melbourne, Melbourne, Victoria 3010, Australia.*

Preface

Water, when it can, flows downhill. From this simple fact follow a host of ecological and (as this volume reminds us) social consequences. One of these is flooding. To us floods may be disasters, but to creatures that live in, on, or around rivers they are a natural, unavoidable, and even beneficial aspect of the environment.

Floods are predictable in the large and unpredictable in the small. The eventual occurrence of a flood at a given spot may be an event of very high probability, and any species (including our own) living there will do well to anticipate it. On the other hand, whether a flood will occur *this* year, and if so what its properties will be, can be as unpredictable as the weather (on which it may depend). This combination of predictability at the statistical level and unpredictability at the event level makes floods an archetypal example of a stochastic environmental factor (along with physical disturbance in the inter-tidal and fire in grasslands, to cite only a couple of ecologically well-studied examples). This volume explores floods as an environmental factor in the context of the Murray–Darling Basin in Australia. It considers the climate and physical background, ecological effects of and adaptations to floods, impacts on the population, landscape, and societal levels, and effects seen through the eyes of birds, pathogens, and economies.

The value of such an integrative treatment can be seen from a perspective—formal demography—that is not emphasized in the book. The dynamics of populations in stochastic environments depends on time variation in the vital rates (a catchall term for rates of survival, reproduction, growth, and so on), which in turn produces a time-varying projection of the population. Over a long time, the statistical properties of the vital rates and of the population structure give rise to a long-term average growth rate. This growth rate determines the long-term fate of the population, measures the fitness of life histories, and provides a tool for analyzing management strategies.

For example, *Boltonia decurrens* is a threatened herbaceous plant restricted to the flood plain of the Illinois River in the United States (Smith *et al.*, 2005). Its life cycle is tightly linked to seasons whose operational boundaries are defined by flooding and flood recession. The timing of flood recession has a major demographic impact. Years with early flood recession lead to explosive population growth. Years with late flood recession lead to precipitous population decline. A stochastic demographic analysis showed that positive long-term population growth can occur only if the frequency of years with early-receding floods exceeds a critical threshold. Part of the success of *Boltonia* was due to the flexibility of its life cycle in which the contributions of annual, biennial, and clonal perennial pathways shift depending on the flood condition (Smith *et al.*, 2005).

Boltonia is not successful now because, in Illinois, the timing of floods (and many other aspects of the hydrology) has changed dramatically over the last century due to the construction of dams and levees. The consequences have been dire; the stochastic growth rate has been reduced below replacement level. Not because of habitat destruction *per se* (although there has certainly been a lot of that) but because of a change in the statistical properties of a stochastic flood environment.

Such changes can come from many sources, as this book emphasizes. One is climate change. Others reflect economic and social forces influencing human impacts on river systems. The conservation implications of the stochastic demography of any threatened floodplain species are played out in the context of all these forces. The editor and authors of this volume are to be congratulated for providing such a wide-ranging view of all these issues.

REFERENCE

Smith, M., Caswell, H. and Mettler-Cherry, P. (2005). Stochastic flood and precipitation regimes and the population dynamics of a threatened floodplain plant. *Ecol. Appl.* **15**, 1036–1052.

Hal Caswell
Biology Department
Woods Hole Oceanographic Institution
Woods Hole, Massachusetts

Acknowledgments

I am extremely grateful to the Murray–Darling Association, especially Ted Lawton for suggesting the commemoration of the Murray–Darling Basin 1956 floods that inspired the production of this book and Brian Grogan for his contagious enthusiasm and commitment.

The multiple and sometimes complex causes and effects of floods cannot possibly be understood without a multidisciplinary approach. I am extremely grateful to all contributors who came on board in this collective effort and did a wonderful job under trying environmental conditions, as, on top of their already busy schedules, they also had to endure a deluge of annoying e-mails reminding them of approaching deadlines.

Many thanks to both Kirsten Funk and Andy Richford from Elsevier for believing in this project and for overseeing the production process with great efficiency and to Hal Caswell, Editor of the Advances in Ecological Research series for his support and for contributing the Preface.

I am grateful to the La Trobe University Publications Committee for granting a subsidy for the publication of color figures.

Last but not the least, I have to thank my wife Marisa and my daughter Catiray for feeding my mind with their love when overcommitment to work menaced with starving it.

Aldo Poiani
Faculty of Science, Technology and Engineering
La Trobe University
Mildura, Victoria
Australia

To my wife Marisa and my parents Franco and Vanda

Contents

Introduction

ALDO POIANI

Climatic Background to Past and Future Floods in Australia

BARRIE PITTOCK, DEBBIE ABBS, RAMASAMY SUPPIAH AND ROGER JONES

Floods Down Rivers: From Damaging to Replenishing Forces

SAM LAKE, NICK BOND AND PAUL REICH

Effects of Floods on Distribution and Reproduction of Aquatic Birds

ALDO POIANI

The Landscape Context of Flooding in the Murray–Darling Basin

ANDREA BALLINGER AND RALPH MAC NALLY

Effect of Flooding on the Occurrence of Infectious Disease

COLIN R. WILKS, ANDREW J. TURNER AND JOSEPH AZUOLAS

Some Economics of Floods

JOHN FREEBAIRN

Tiddalik's Travels: The Making and Remaking of an Aboriginal Flood Myth

JOHN MORTON

Understanding the Social Impacts of Floods in Southeastern Australia

CATHERINE ALLAN, ALLAN CURTIS AND NICKI MAZUR

Introduction

ALDO POIANI

1. *Fresh water running, splashing, swirling,*
    Running over slippery stones ... clear water ...
    Carrying leaves and bushes before it ...
    Swirling around ...

2. *Water running, running from pool to pool ...*
    Water running in streams,
    Foaming, carrying leaves and bushes before it ... churning,
    Bubbling up among the Miljarwi clansfolk.
    Water flowing over the rocks ... flowing each side of the termite mounds,
    Running fast toward Nalibinunggu clansfolk ... Ridarngu ... Gaiilindjil ... Ridarngu,
    Toward the Bunangaidjini Wonguri ...
    Fast-running water.

3. Bilgawilgajun! *(We invoke the Spirits!)*
    Water dammed up by barriers of stone at Buruwandji ... at
    Mumana, at Bungarindji,
    Breaking out, foaming, like sacred feathered armbands ...
    Carrying away the debris ...
    Sound of rushing water ... running,
    Smaller streams joining together, roaring down ...
    Washing out tree roots at Buruwandji ... running past the Rocks.
                    Ridarngu song from Arnhem Land (Berndt and Berndt, 1977, pp. 375–376)

I. FLOODS IN AN ARID CONTINENT

On the morning of July 7, 1956, the town of Mildura in northwestern Victoria, Australia, a usually tranquil city overlooking the Murray River woke up to the news in the local newspaper, the *Sunraysia Daily*, that "Floodwaters are pushing hard against the network of levee banks around

ADVANCES IN ECOLOGICAL RESEARCH VOL. 39
© 2006 Elsevier Ltd. All rights reserved

Wentworth and should they break through the town will be flooded." Two days later, on July 9, the headlines announced that "Roads cut as floods spread in 3 states," while "help of army sought to hold weakened levee bank." The situation continued to worsen in the following days and the issue of July 11 warned on page 1 of a "Desperate fight to hold river. Thousands of people along the banks of the flood-swollen Murray and its tributaries are fighting desperately to save their homes," and on page 2, the people of Mildura began to realize the magnitude of the unfolding disaster as the "District faces worst flooding ever known—1931 peak neared." Two days later, the eyes were on the levees as the flood level was expected to "near peak," and on July 24 the paper grimly announced that "The main levee bank on the Murray at Mildura broke yesterday to make the present flood the most costly in the district's history." On July 26, the Murray River was "at 1931 level—and still rising." In fact, what the inhabitants of Mildura and Wentworth were experiencing firsthand was the unfolding of the worst flood to involve the Murray and Darling rivers in the twentieth century (Fig. 1).

Floods, generally defined as "a significant rise of water level in a stream, lake, reservoir or coastal region" (IRIN, 2005), have gained a dual identity of sorts. On the one hand, daily news abound in detailed accounts of natural disasters, with spectacular floods being a favorite subject of media attention. I am writing this chapter in mid-April 2006 and a quick survey of the Australian Broadcasting Corporation, the American Broadcasting Corporation, CNN, and BBC Web sites for the first three and a half months of the year uncovers news about floods in Australia (Katherine River, Northern Territory; Murchison catchment, Western Australia; Bellinger River, New South Wales), Indonesia, India, United States (North Dakota, Minnesota), Malawi, and several countries in Europe (Czech Republic, Germany, Bulgaria, Greece, Poland, Hungary, Austria, Slovakia). Not surprisingly, floods are considered a major "natural disaster" that should be prevented or its effects on human populations mitigated if possible. On the other hand, floods can distribute nutrients throughout vast areas of floodplains and contribute to the maintenance of both natural and some agricultural ecosystems.

Floods have been impacting on biota, including human populations, since time immemorial. Plant and animal species inhabiting floodplains show adaptations presumably selected in response to recurrence of floods in the environment (Lytle and Poff, 2004). For instance, *Eucalyptus camaldulensis* is a typical flood tolerant tree species in Australia that has several physiological adaptations allowing trees to withstand the effects of flooding (Blake and Reid, 1981). From the particular perspective of our species, ancient civilizations developing along the basin of major river systems in all continents have been affected in one way or another by floods (Killick, 1985; Shendge, 1991; Hassan, 1998). Human experience with floods has given rise

Figure 1   Murray–Darling flood of 1956. The 1956 flood was the most intense in the Murray–Darling Basin in the twentieth century. © Ian Hehir.

not only to flood myths and stories (e.g., see the Ridarngu song quoted above) but also to practical solutions to either exploit the benefits of floods or prevent the negative effects of floods through the establishment of flood mitigation procedures.

Australia is widely recognized as a mainly dry continent, with about 70% of the mainland being categorized as arid or semiarid (Stafford-Smith and Morton, 1990). Yet, the risk of floods in the country is high in some regions. Although this book mainly focuses on the Murray–Darling Basin (MDB), a major system of rivers, wetlands, and floodplains covering a large extension of southeastern Australia, other areas at high flood risk, especially in the tropical north that is under a monsoonal regime, will be also considered in some of the chapters. Moreover, the MDB is also subject to the effects of climate in the tropics as floods in the Darling River are affected by the tropical climatic conditions that characterize the northern areas of the continent.

The book is organized into nine chapters, including this introductory chapter, that bring together experts in climatology, limnology, ornithology, landscape ecology, epidemiology, economics, anthropology, and sociology, addressing the causes and consequences of floods in Australia from the perspective of the physical environment, the biota, and human populations. In what follows, I will introduce the different chapters first to subsequently propose a preliminary synthesis of our knowledge of floods in Australia, with the aim of drawing some general conclusions that may be applicable to other regions as well.

Pittock *et al.* in their chapter review the *Climatic background to past and future floods in Australia*. They describe how the current climate in the northern parts of the Darling River catchment, especially in summer, is under the influence of troughs in the low-latitude easterlies, that is, winds moving from east to west, and tropical atmospheric pressure lows that affect the extent of rainfall in the tropical rainy season; whereas rainfall in the southern regions of the MDB is affected by winter atmospheric pressure lows and also cold temperatures resulting from winds coming from the west (westerlies). These climatic conditions affecting the MDB determine the occurrence of summer floods on the Darling River and mostly winter floods along the Murray River. A major climatic influence on the probability of floods in the MDB is the El Niño-Southern Oscillation (ENSO). In the opposite phase of ENSO, not surprisingly called "La Niña," the northern and eastern parts of Australia tend to be subject to high rainfall that may cause floods. However, as indicated by Pittock *et al.*, ENSO is not the only source of interannual variability in flooding as midlatitude westerly winds may also be relevant in the southern regions of Australia.

Instrumental records of past climatic conditions suggest not only fluctuations in occurrence of floods associated with ENSO but also long-term trends correlated with global warming. Global climate change is bound to affect the probability of flooding in Australia. Whereas the Northern Hemisphere has shown trends toward increased continental precipitations in the twentieth century, the Southern Hemisphere has shown a tendency to an increase in frequency and persistence of El Niño events (Watson and Core Writing Team, 2001). Chapter by Pittock *et al.* also offers interesting projections on future climate change based on the most recent Australian Commonwealth Scientific and Research Organization (CSIRO) analyses. According to these results, the MDB is expected to be subject to lower winter and spring rainfall by the year 2070 along with increases in summer rainfall of up to 40%. However, as suggested by Pittock *et al.*, projections of this kind also come with a healthy note of caution, especially because it is difficult to model the influence of ENSO on precipitations if global temperature continues its raising trend.

In chapter *Floods down rivers: from damaging to replenishing forces*, Sam Lake, Nick Bond, and Paul Reich compare flood dynamics and their ecological effects on both upland and lowland ecosystems. Floods are seen as pulse disturbances that usually vary in character—depending on whether we are focusing on upland or floodplain regions—to which biota may be resistant or resilient. In constrained upland water courses, floods are associated with fast moving water that carries sediments, debris, and organisms, whereas in unconstrained floodplain rivers floodwater has usually less velocity but it has the capacity to inundate vast areas. Organisms have evolved adaptations in response to the specific environmental challenges posed by different kinds of floods. Lake *et al.* suggest that in upland streams biota face high energy but short-durational water flows during floods to which the selection of traits, such as small body size, high adult mobility, substrate-clinging, and others, may be favored. On the other hand, the biota living in floodplains are likely to evolve strategies to both maximize the use of resources available during a flood and escaping the effects of desiccation when floodwater recedes.

The chapter concludes with a section addressing the effects of human interventions that have altered flood dynamics in the MDB. Damming of streams results in ecological impacts on upland sections of the MDB, whereas lowland floodplains have been adversely affected by river flow regulation and water extraction. Reestablishing some degree of natural water flows that allow controlled flooding to take place is an urgent measure to be taken if we want to redress the damage caused by ecologically inappropriate water regulation policies. Lake *et al.*'s plea for more research in this area carries with it a considerable degree of urgency.

Aldo Poiani's chapter reviews *The effects of floods on distribution and reproduction of aquatic birds*. This chapter has a broader, continental-wide framework, emphasizing case studies from both the MDB and also the tropical and arid regions of Australia. Floods are environmental factors that can affect life histories of aquatic birds. Long-living species with high capability for long-distance movements and an opportunistic capacity to reproduce are best suited to survive and reproduce in an environment subject to unpredictable floods. In general, birds respond to a flood by moving out of the flooded area in the first instance to then take advantage of the nutrients carried by floods to reproduce, to finally concentrate in great numbers at remaining water bodies as the floodwaters start to recede.

The spectacular concentration of aquatic birds that regularly occurs in flooded endorheic lakes in the Australian arid zone is likely a result of an adaptation of birds to exploit the availability of large concentrations of food for survival and reproduction in an environment relatively safe from predators. However, the mechanisms used by aquatic birds to track down the location of flooded inland lakes from a distance of hundreds of kilometers remain largely unknown. Here Poiani proposes a list of potential mechanisms to explain how could aquatic birds, which inhabit the coastal regions of the continent, detect the availability of flooded endorheic lakes. A combination of sensory abilities, such as detecting atmospheric infrasounds and gradients of volatile chemicals, may be able to explain the ability of birds to detect the long-distance occurrence of floods and to track the exact location of the flooded area.

Chapter by Aldo Poiani also suggests that flood regulation, for example along the MDB, may negatively affect viability of aquatic bird populations. Flood regulation strategies compounded by ENSO and global climatic change may have especially detrimental effects on aquatic bird populations in southeastern Australia.

Andrea Ballinger and Ralph Mac Nally in their chapter analyze *The landscape context of flooding in the Murray–Darling Basin*. This chapter mainly focuses on the effects floods have on floodplain ecology and it uses Junk *et al.*'s (1989) "flood pulse concept" and a more general and synthetic model proposed by Walker *et al.* (1995), as central theoretical frameworks. In brief, the flood pulse regime determines a degree of environmental variability and sometimes unpredictability that may tend to select for opportunistic and flexible life-history strategies in many populations of organisms. At the same time, floods increase connectivity of river–floodplain systems and help dispersal processes of some species. When those species (e.g., arthropods) are food to other species (e.g., birds and bats) then the effects of floods may reverberate throughout the floodplain affecting many terrestrial biota. Ballinger and Mac Nally identify three major strategies that could be selected in organisms to cope with the unpredictability of flooding: (i) Switch to reproductive mode in response to any flood (e.g., *E. camaldulensis*), (ii) Switch to reproductive mode only when a threshold flood level is reached (e.g., many aquatic birds), and (iii) Switch to reproductive mode following cues other than floods (e.g., many fish).

Chapter by Andrea Ballinger and Ralph Mac Nally emphasizes the effect of floods on the enhancement of landscape connectivity (both longitudinal and lateral). Such connectivity is important for the maintenance of biodiversity in both aquatic and terrestrial environments along floodplains, and it should be carefully taken into account in the process of designing and implementing flood regulation policies.

Colin Wilks, Andrew Turner, and Joseph Azuolas in their chapter analyze the *Effect of flooding on the occurrence of infectious disease*. Floods can have serious consequences on animal and human populations in terms of favoring the conditions for disease outbreaks. Both waterborne and some arthropod-borne disease agents can be more easily transmitted under flooding conditions (Connolly *et al.*, 2004). Wilks *et al.* review our current knowledge of some micropathogens (mainly viruses, but also *Bacillus anthracis*, a bacterium) that show very diverse levels of ability to undergo outbreaks following floods. Of the micropathogens included in this chapter, those transmitted by arthropod vectors, such as flaviviruses (e.g., *Murray Valley encephalitis virus*) and alphaviruses (e.g., *Ross River virus*), are clearly the most sensitive to flooding, as floodwaters produce optimal breeding conditions for the mosquito vectors. Human and animal hosts can be exposed to arbovirus (i.e., arthropod-borne virus) infection whenever they share the same environment with mosquitos at the same time when mosquitos are active. Wilks *et al.* describe the development of a recent scenario in the MDB where different mosquito species, that is, *Culex australicus, C. annulirostris,* and *Ochlerotatus camptorhynchus* replace each other as potential vectors to pathogenic arboviruses in the same area as they have different environmental constraints, thus producing a continuous risk of exposure through time in spring, summer, and autumn. The case of *O. camptorhynchus* is especially interesting as this is a species that not only transmits *Ross River virus* but it also requires salty water for reproduction, suggesting that the current trend toward increasing salinity in many areas of the MDB (Newman and Goss, 2000) may favor future outbreaks of the disease especially in co-occurrence with heavy floods. Most of the mosquito species are endemic to the tropical areas of Australia, thus it is expected that if global climate changes increase precipitations in the tropics producing floods in some southern regions (e.g., in the Darling River), then mosquito populations may increase over time in some areas of the MDB thus favoring transmission of viral diseases to both humans and animals.

Some economics of floods is the focus of John Freebairn's chapter. With world economic losses from weather-related events having increased almost 20-fold from the 1950s to the 1990s (IRIN, 2005), major environmental hazards, such as floods, are likely to condition not only the allocation of resources for natural disaster prevention but also resources to be used in alleviating the effects of floods once those floods have occurred and, ultimately, the decision on whether a flood-prone area should be inhabited in the first place. Such decisions are made at all levels in society from the individual to local and national governments. When it comes to allocating economic resources, Freebairn provides us with the healthy reminder that decisions are more often than not made under the constraint of limited resources, thus we are usually faced with the need to make tough choices.

Whether to locate a household, a town, or an economic activity such as a farm in any area that may be subject to flooding should be a decision guided by properly informed cost–benefit analysis. One of Freebairn's central tenets is that when they are based on maximization of net benefits and, importantly, full disclosure of information regarding flood event frequency and severity, individual choices will approximate the society choices if there are not market failures. This, however, does not preclude society to absorb some of the costs of activities carried out in flood-prone areas when overall benefits outweigh costs (e.g., in situations where land is scarce). Some such costs are those involved in flood mitigation initiatives, including the ecological costs of flood mitigation. Although chapter by John Freebairn uses a cost–benefit analytical approach, it does mention the availability of alternative models where some costs (e.g., to the environment) are not given a dollar value.

The sensitive issue of subsidies is discussed at length by Freebairn. Subsidies provided to households or businesses to establish themselves in flood-prone areas are not only debatable in economic terms, as it is done in this chapter, but also from an environmental point of view as the resources spent in the subsidy could have been diverted to, for instance, environmental protection or restoration of floodplains following the alternative relocation of those households or businesses to less flood-prone areas.

John Morton in his chapter provides an anthropological perspective with his analysis of *Tiddalik's travels: The making and remaking of an aboriginal flood myth*. Given that floods are phenomena of common occurrence in many regions of the world, it is not surprising that flood myths and stories are also common and widespread, with Isaak (2002) providing a collection of over 350 flood stories from all continents, including several Aboriginal stories from all climatic regions in Australia. Morton, however, analyzes one specific story—that of Tiddalick the Frog—in great detail uncovering the links between environmental phenomena and culture in Aboriginal perspective.

The chapter by John Morton introduces us into the "Dreamtime" world of Aboriginal cosmology where all parts of that world are both autonomous and in relationship to each other. The relationship is maintained in balance only when the parts are of equal "power," and the balance can be achieved if those parts communicate with each other. Thus, balance in the Aboriginal world is not a natural state of affairs, rather it is a state that the parts may reach starting from situations of disharmony. In Morton's words and from an Aboriginal perspective, a flood occurs *when the world gets "out of kilter."*

In Tiddalick's story, a period of drought is broken by a flood event as soon as the water held in the frog's mouth is released following the frog burst into laughing after watching Eel perform a contorting dance. In a land where extremes of climate (e.g., droughts versus floods) are common, Aboriginal cosmology and social organization reflect to some extent such variability through the "moieties" organization of social groups. Moieties are identifiable social units that however interact with each other (e.g., through cross-moiety marriage). Moieties do not establish harmonious relationships with each other as a matter of fact, but as a result of an active process where both obligations and demands are met.

In Tiddalick story, Morton suggests that floods are also represented in a duality fashion, as *bringer of both life and death.*

A sociological perspective on floods is provided by Catherine Allan, Allan Curtis, and Nicki Mazur in their chapter *Understanding the social impacts of floods in southeastern Australia*. Flood risk is regarded as the most severe environmental hazard in the Asia-Pacific region (IFRC, 2000), and floods are also the most frequently reported natural disasters in Africa, Asia, and Europe (IRIN, 2005). Allan *et al.* carry out their analysis within the framework of risk perception and social impact assessment (SIA) theory that considers the manners in which particular events (e.g., a flood) and policies will affect human communities and the parameters (monetary and not) associated with their mode of life. The chapter focuses on four of the seven steps of the SIA model: identification of impacts of floods (scoping), evaluation of those impacts (evaluation), set priorities for strategies that mitigate some of those impacts (mitigation), and development of a monitoring program (monitoring). The model is applied to the analysis of three case studies drawn from floods and flood regulation schemes experienced in the state of Victoria, southeastern Australia. Allan *et al.* draw several lessons from the analysis of those case studies. Winners and losers resulting from floods and flood control interventions are predictable, along with the identity of individuals and communities at risk, consequently it is possible to act plans aimed at minimizing negative effects of floods on humans. This process must be inclusive, participatory, and informative, with local communities being fully aware of relevant knowledge acquired by experts in different fields of research.

II. CAUSES AND CONSEQUENCES OF FLOODS IN AUSTRALIA: A POSSIBLE SCENARIO

Floods in Australia affect various regions of the continent, including arid and semiarid zones. If predictions derived from current climatic models are correct, precipitations in the northern, tropical areas of Australia may be in the increase. This may not only change the landscape of the central arid zone, as floodwaters may drain into currently endorheic lakes more frequently and regularly, but it may also affect southern temperate regions such as the MDB through flooding of rivers such as the Darling. If El Niño years become more frequent however, the frequency of flooding events in the southern parts of the basin may decrease. Such a potential scenario may set the stage for substantial changes in the ecology of the continent which will also condition many human activities. Although medium and long-term climatic predictions are always subject to a degree of error, continuous monitoring of climatic effects on biota and the establishment of rapid mechanisms of response to address local problems before they become unmanageable are what we need to provide us with the flexibility of action needed in the face of uncertainty.

If recurrence of floods of tropical origin increases availability of permanent water bodies in internal regions of Australia and the northern areas of the MDB, this may not only increase dispersal and population growth of aquatic organisms, such as fish and invertebrates, but also the distribution and population sizes of aquatic birds and terrestrial species living in and around floodplains. Increased connectivity of the landscape caused by floods may also allow the availability of permanent north-south corridors for the spread of aquatic species and also pathogens such as mosquito-borne viruses from the tropics to the MDB. This process may be of particular human concern in the case of pathogenic flaviviruses.

If the southern parts of the MDB are likely to become drier in the future, then the current floods regulation and water usage regimes should be carefully reconsidered if ecosystem sustainability and, consequently, long-term sustainability of human communities are to be achieved.

This book is published on the occasion of the 50th anniversary of the 1956 floods that ravaged parts of the MDB. Since then, the major rivers in the MDB have been subject to flow regulation and the water is used in a significant manner for human activities, especially agriculture. In recent times, our perception of floods in the MDB has slowly shifted from *natural disaster* to *ecological necessity*; I hope that after reading the contributions collated in this book, we may realize that, in many ways, they are both.

ACKNOWLEDGMENTS

I am grateful to Ros Gerrard from Aboriginal Studies Press for granting permission to reproduce the Ridarngu song and to Ian Hehir for granting permission to reproduce Fig. 1.

REFERENCES

Berndt, R.M. and Berndt, C.H. (1977) *The World of the First Australians*. Ure Smith, Sydney.

Blake, T.J. and Reid, D.M. (1981) Ethylene, water relations and tolerance to waterlogging of three *Eucalyptus* species. *Aust. J. Plant Physiol.* **8**, 497–505.

Connolly, M.A., Gayer, M., Ryan, M.J., Spiegel, P., Salama, P. and Heymann, D.L. (2004) Communicable diseases in complex emergencies: Impact and challenges. *Lancet* **364**, 1974–1983.

Hassan, F.A. (1998) Climatic change, Nile floods and civilization. *Nat. Resour.* **34**, 34–40.

IFRC (2000) "World Disasters Report 2000." International Federation of Red Cross and Red Crescent Societies, Geneva.

IRIN (2005) "Disaster Reduction and the Human Cost of Disaster." Integrated Regional Information Networks, UN Office for the Coordination of Humanitarian Affairs, New York.

Isaak, M. (2002) Flood stories from around the world. http://home.earthlink.net/~misaak/floods.htm

Junk, W.J., Bayley, P.B. and Sparks, R.E. (1989) The flood pulse concept in river-floodplain systems. *Can. J. Fish. Aquat. Sci.* **106**, 110–127.

Killick, R.G. (1985) Flood control in northern Babylonia. *Sumer-Baghdad* **41**, 121–123.

Lytle, D.A. and Poff, N.L. (2004) Adaptation to natural flow regimes. *Trends Ecol. Evol.* **19**, 94–100.

Newman, B. and Goss, K. (2000) The Murray-Darling Basin salinity management strategy—implications for the irrigation sector. Paper presented at Annual ANCID Conference, Toowoomba, Australia.

Shendge, M.J. (1991) Floods and the decline of the Indus civilization. *Ann. Bhandarkar Orien. Res. Inst.* **71**, 219–263.

Stafford-Smith, D.M. and Morton, S.R. (1990) A framework for the ecology of arid Australia. *J. Arid Environ.* **18**, 255–278.

Walker, K.F., Sheldon, F. and Puckridge, J.T. (1995) A perspective on dryland river ecosystems. *Regul. Rivers Res. Manag.* **11**, 85–104.

Watson, R.T. and Core Writing Team (Eds.) (2001) "Climate Change 2001: Synthesis Report. A Contribution of Working Groups I, II, and III to the Third Assessment Report of the Intergovernmental Panel on Climate Change." Cambridge University Press, Cambridge.

Climatic Background to Past and Future Floods in Australia

BARRIE PITTOCK, DEBBIE ABBS,
RAMASAMY SUPPIAH AND ROGER JONES

SUMMARY

Rainfall variability in Australia is generally among the highest in the world, largely due to the dominant influence of the El Niño–Southern Oscillation. Australian climate has been characterized by wet and dry periods with sometimes sudden transitions from one mode to the other. Synoptic explanations and teleconnections are discussed, with an emphasis on the Murray–Darling Basin (MDB). Floods in Australia are generally of two types: local "flash floods" and widespread basin flooding. The latter often start in the upper reaches of the MDB due to tropical low-pressure systems, and slowly move downstream. Floods in smaller subcatchments can be rapid and intense due to heavy thunderstorms or other instability exacerbated by topography.

Global warming due to human activities is occurring. Observed rainfall trends may be partly natural but partly of human origin. Possible changes to rain-bearing systems associated with tropical and midlatitude systems will be discussed, along with land cover change and possible effects on runoff, erosion, and sediment loading. The general picture for the MDB, despite significant uncertainties, is that the frequency of small catchment flash flooding is likely to increase, especially in autumn and winter and in summer

ADVANCES IN ECOLOGICAL RESEARCH VOL. 39
© 2006 Elsevier Ltd. All rights reserved

0065-2504/06 $35.00
DOI: 10.1016/S0065-2504(06)39002-2

over the northern and southern (but not central) areas of the MDB. Large catchment floods may increase in magnitude in summer in the northern parts of the Basin, but are less likely to do so in the south, where average rainfall is likely to decrease. Increased fire risk and decreased ground cover may lead to greater sediment loading and erosion during floods, although reforestation could slow runoff and decrease magnitudes of floods, especially in the upper catchments.

I. GEOGRAPHIC AND CLIMATIC SETTING

Humans tend to see floods as water that is out of place, inundating normally dry land used or potentially used by humans. However, floods can also be seen as part of the natural process that deposits silt and nourishment on flood plains and periodically maintains vegetation and wildlife, especially water birds. Floods as a hazard are largely conditioned by the degree of human adaptation to floods. In Australia's early years of European colonization, flood hazards loomed large because of lack of knowledge of the magnitude of infrequent floods and the consequent occupation of unsuitable low-lying areas.

Subsequent siting of human development on higher ground, the building of dams and the provision of flood diversion channels and levees, has greatly reduced flood hazards. Even so, the estimated average annual cost of floods in Australia in the years 1967–1999 was $314 million, exceeding that from severe storms ($284 million), cyclones ($266 million), and bushfires ($77 million) (BTE, 2001).

However, if floods become more frequent and of greater magnitude with climate change, the frequency, severity, and damage from floods could again increase.

The magnitude and frequency of floods depend on rainfall amount and intensity, but they are significantly affected by the characteristics and condition of the catchment, including ground cover and soil moisture. Moreover, intense rain and runoff in sparsely covered catchments lead to sheet, rill, and bank erosion and high sediment loadings. This is likely to reduce water quality and deposit large quantities of sediment (mainly sand) in river beds, pools, and reservoirs, changing the ecology, channels, and flow rates of rivers. The history of past flood regimes can be pieced together from the relics of these changes.

The Murray–Darling Basin (MDB) occupies an area of more than a million square kilometers in southeastern Australia. It has an average annual runoff of about 24,000 GL/year, of which only some 13,000 GL/year flows into the sea under natural conditions, the rest being consumed by wetlands and floodplains. Average annual rainfall varies from less than 300 mm/year in the west to in excess of 1000 mm/year in the southeastern highlands, with rainfall in excess of 600 mm/year in most of the eastern fringes to the west of the Great Dividing Range as far as 27°S. Rainfall intensity tends to be greatest across northern Australia due to tropical influences, and down the east coast, mainly east of the Great Dividing Range. Peak catchment discharges within the MDB are generally less than 1 $m^3/(s\ km^2)$, which is a factor of 10 or more less than in the smaller east coast catchments. The low discharge rates in the Basin are largely because of lower rainfall intensity, the relative flatness of the catchments, and the ability of the soils to soak up large amounts of water, which becomes groundwater.

Nevertheless, large area heavy rainfalls, often lasting several days, can occasionally lead to widespread flooding, while intense thunderstorms in the hillier upper catchments can lead to short-lived floods in smaller areas. An example of the former occurred in April and May of 1983, with drought-breaking rain of 188 mm over 11 days at Goondiwindi, leading to a flood peak of 10 m and widespread stock losses. This flood peak took weeks to travel down the Macintyre and Barwon Rivers. A notable flash flood in the Woden Valley in Canberra in 1971 caused major property damage and the loss of seven lives and led to new flood control works. A survey by Devin and Purcell (1983) found some 88 settlements of over 200 people that were at risk from a once in 100-year flood in the MDB, including 11,620 buildings.

Land-clearing and water diversions for irrigation have greatly changed the Basin hydrology, leading to faster runoff and greater soil erosion, but greatly reduced average river flow and frequency of flooding (Thoms et al., 2004), with adverse effects on many wetlands (Kingsford, 2000). Surface water extraction is now about 12,000 GL/year on average, with annual median flows to the sea down to about 3000 GL (Craik, 2005). Major water storages include the Hume Reservoir (now 3038-GL capacity), built in 1936, and the Dartmouth Reservoir, completed in 1979 of 3906-GL capacity. Under high flow conditions water is also stored in the natural but usually dry Menindee Lakes on the Darling River below Wilcannia [New South Wales (NSW)], and at Lake Victoria below Wentworth (NSW) on the Murray River. Principal locations are shown in Fig. 1.

The extent of reductions in flood flows due to human development is well illustrated in Table 1, which shows reductions in simulated flood discharges relative to natural conditions in the Darling River. In individual major flood events (return interval greater than 20 years) up to some 30% of discharge can be abstracted above Wilcannia.

Major climatic influences affecting rainfall are troughs in the low-latitude easterlies (Annette, 1978), tropical lows (some from tropical cyclones which

THE MURRAY–DARLING BASIN

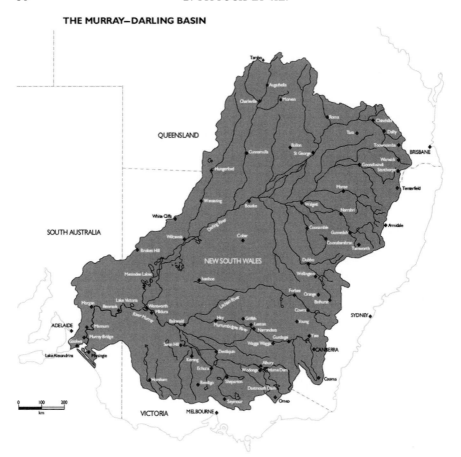

Figure 1   Map of the Murray–Darling Basin (courtesy of the Murray–Darling Basin Commission).

Table 1   Simulated flood discharges (ML/day) for 1993/1994 levels of development, and their percentage change from natural or "reference" conditions for four stations, where ARI is the average recurrence interval

ARI	Mungindi	Walgett	Bourke	Wilcannia
2	1758 (−7%)	5,299 (−40%)	4,737 (−34%)	4,487 (−34%)
5	2120 (−40%)	10,741 (−26%)	9,914 (−22%)	5,270 (−18%)
10	3022 (−21%)	19,598 (−21%)	18,188 (−17%)	10,041 (−17%)
25	3574 (−9%)	41,167 (−25%)	32,610 (−17%)	28,863 (−36%)

Based on Table 6 of Thoms *et al.* (2004).

have moved inland) in the northern part of the Darling catchment in summer; northwest and northern oceanic cloud bands (Wright, 1997), cutoff and east coast lows (Holland *et al.*, 1987) affecting the eastern headwaters, spillover from the trade winds in the northeast, and winter lows and cold fronts embedded in the westerlies (Wright, 1988a) in the southern regions. Discussions of synoptic situations associated with some of these phenomena will be found in Tapper and Hurry (1993) and in Abbs *et al.* (2005).

The seasonality of these rains is such that high runoff events occur almost exclusively in the summer half year in the Darling catchment (Moore, 1988), and mostly in winter in the more southern Murray catchment, although small subcatchments can experience high runoffs due to localized summer storms in the southern regions (Grimmer, 1988).

II. INTERANNUAL VARIABILITY OF RAINFALL

The large variability of Australian rainfall is due to variations in large-scale influences including the El Niño–Southern Oscillation (ENSO), the latitude of the westerlies and variations in standing wave patterns associated with blocking phenomena (Fig. 2).

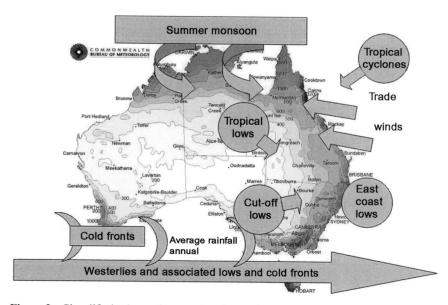

Figure 2 Simplified schematic map showing major synoptic influences on rainfall in the MDB (base map courtesy of the Bureau of Meteorology). (See Color Plate 1.)

The pattern of rainfall variability associated with ENSO is a dominant influence on the frequency and intensity of floods in the MDB, with the center of its influence situated right over the Basin, where it accounts for up to a third of the total interannual variance of rainfall (Troup, 1965; Pittock, 1975). In La Niña years, the surface pressure difference between the quasi-stationary high-pressure region in the southeastern Pacific Ocean and the quasi-stationary low in the Indonesian region is large. In such years, rainfall over northern and eastern Australia tends to be high, with a strengthened Australian summer monsoon, a greater incidence of tropical cyclones off the northeast coast, and a much greater likelihood of floods. In El Niño years, the situation is reversed with drought more prevalent over eastern Australia. The relationship between ENSO and Australian rainfall, however, varies with time, and this may be related to wider global influences (Cai *et al.*, 2001; see also Wright, 1988b).

A second major pattern of rainfall variability is associated with the movement north or south of the midlatitude westerly winds and associated low-pressure systems and cold fronts that bring rain to southern Australia (Pittock, 1975; Trenberth, 1976). These movements are associated with what is more recently termed the Southern Annular Mode (SAM) (Thompson and Wallace, 2000; Thompson *et al.*, 2000; Cai *et al.*, 2003; Marshall, 2003). When the SAM is more negative, the westerlies are further north and more winter rain falls in the southern part of the MDB (Meneghini *et al.*, 2006).

III. THE PALEOCLIMATIC RECORD

At the time of the last glacial maximum, some 20,000 years ago, the MDB region was in general wetter, but most of this was due to the much colder conditions, and rainfall was probably less, with fewer intense rain events (Hesse *et al.*, 2004). There is evidence that the Willandra Lakes and others in the region were full and fresh, with Aboriginal occupation and remains nearby (Bowler, 1998; Smith, 2005).

However, by the time of the postglacial maximum warming around 8000–6000 years ago, the Willandra Lakes were dry, and there were high lake levels in Lake George which is now usually dry (Singh and Geissler, 1985; Bowler, 1998; Hope *et al.*, 2004). However, it should be noted that analogies with the present rapid anthropogenic warming (Pittock and Salinger, 1982) may be inappropriate. This is due to the fact that the postglacial warming occurred over thousands of years, allowing the temperatures and circulations of the surrounding oceans to keep pace with changes in radiative forcing. In contrast, the present warming due to increased greenhouse gases is at least 10 times faster, leading to possibly large lags in ocean temperatures and circulations.

Paleoclimatic fluctuations in the strength and latitude of the westerlies is discussed in Shulmeister *et al.* (2004), but interpretation of the evidence is confusing due to the role of humans in the burning of landscapes since some 50,000–40,000 years ago. This confuses the paleoclimatic interpretation of the pollen records (Miller *et al.*, 2005).

IV. THE INSTRUMENTAL RECORD

Instrumental records of Australian rainfall began in the mid-nineteenth century. As records became longer, fluctuations in the rainfall record were analyzed (Deacon, 1953; Kraus, 1954; Gentilli, 1971). They all noted a wetter period in the MDB in the late nineteenth century then a drier period in the first half of the twentieth century. Pittock (1975) analyzed twentieth century rainfall records, examining the influence of the high-pressure belt (and by implication, the latitude of the midlatitude westerlies) and ENSO on interannual rainfall patterns. He found a significant rainfall increase across the late 1940s. Kraus (1955, 1963) related rainfall changes in Australia to similar changes in other countries, suggesting that there may be some global-scale regime changes.

A high-quality rainfall data set for Australia (Lavery *et al.*, 1992, 1997) was analyzed from 1910 by Nicholls and Lavery (1992). They found no significant changes in mean rainfall, although the influence of ENSO and a changed relationship between rainfall and temperature since 1976 was noted by Nicholls *et al.* (1996). Suppiah (2004) has also noted changes with time in the relationship between ENSO and Australian rainfall. Using the same data set, Suppiah and Hennessy (1998) and Hennessy *et al.* (1999) showed significant increases in extreme rainfall (based on annual values of 95th and 99th percentile daily falls) over the century for most of the continent except in southwest Western Australia (SWWA).

Using records of river floods and flow discharge (Warner, 1987, 1995) eastern Australian rainfall was partitioned into flood-dominated regimes (FDRs) and drought-dominated regimes (DDRs) covering the nineteenth and twentieth centuries. This was rejected by Kirkup *et al.* (1998) on both statistical and conceptual grounds, but Whetton *et al.* (1990) noted an abrupt change in river flows from the Darling River in the MDB in early 1890s that coincided with changes in the Nile and in Northern China. Franks (2002) and Franks and Kuczera (2002) have also investigated the effect of decadal rainfall variability using flood gauge data from New South Wales, which supports Warner's hypothesis and confirms the abrupt nature of changes noted by Whetton *et al.* (1990).

Vives and Jones (2005) used a bivariate test for abrupt changes in rainfall, in contrast to the null hypothesis significance test for changes, which assumes stationarity (Nicholls, 2000), that is, the long-term statistics are unchanging.

If one assumes nonstationarity, both decadal variability and long-term changes may be present. The bivariate test does not distinguish between these possible causes, but it does locate changes that may subsequently be attributed (Hulme *et al.*, 1999).

The Vives and Jones analysis examined all Australian rainfall records of over 60 years in length from 1890 to 1989 to determine where and when abrupt rainfall changes occurred. They found three distinct periods in the record:

- A wet or "flood-dominated" regime shifted into a drier or "drought-dominated" regime around 1895 and persisted until about 1945 (Fig. 4A).
- A flood-dominated regime began in eastern Australia in 1946–1948, and a drought-dominated regime commenced in SWWA at about the same time.
- Further decreases in rainfall in SWWA and increases in eastern Australia occurred during the period 1967–1973.

The cause of such abrupt multidecadal fluctuations is as yet not clear. They could be associated with natural modes of behavior in the climate system, with abrupt switches from one to the other when some trigger or tipping point is reached due to the accumulation of random (stochastic) events. Similar abrupt changes might be brought on by gradual changes in climate forcing such as those from increasing greenhouse gases, so change due to the enhanced greenhouse effect may be gradual or, at times, abrupt.

The long-term persistence of drought- or flood-dominated rainfall regimes, and possible shifts from one to another, will affect the risk of floods (Warner, 1995; Franks, 2002) and droughts (Jones and Pittock, 2002). This means that such persistence needs to be incorporated into any stochastic generation of climatic data to be used in planning and managing catchments (Srikanthan and McMahon, 2001).

V. RECENT TRENDS AND OBSERVATIONS

Observational data indicate that the climate of Australia and the MDB has changed considerably since 1950. Both Smith (2004) and Collins and Jones (personal communication, 2005) have used data from the Bureau of Meteorology high-quality station network to analyze trends in rainfall and temperature for Australia.

Strong contrasts in rainfall trends (Fig. 3A) are evident throughout the country and seasons since 1950. The eastern part of the country, including the MDB, has experienced strong decreases in annual rainfall. Changes in summer rainfall (Fig. 3B) are similar to those for annual rainfall. Autumn rainfall (Fig. 3C) shows the lower MDB mainly recording declines with a

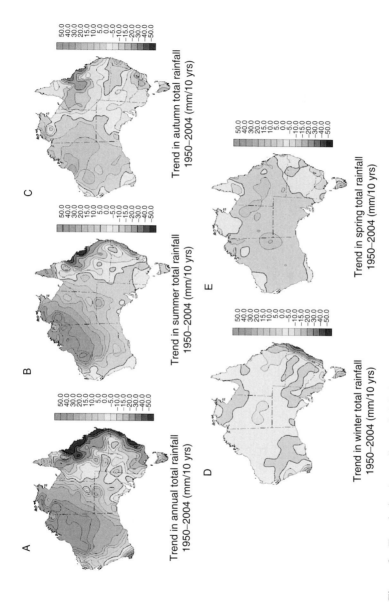

Figure 3 Trends in Australian rainfall from 1950 to 2004. (A) Annual, (B) summer, (C) autumn, (D) winter, and (E) spring (courtesy of Dean Collins, Bureau of Meteorology). (See Color Plate 2.)

weak increase in the northern portion of the catchment. Winter (Fig. 3D) and spring (Fig. 3E) rainfall changes are relatively weak. Rainfall is highly variable and the period over which it is examined strongly influences trend values. The post-1950 decline in annual rainfall over eastern Australia is largely due to some very wet years during the 1950s and some very dry El Niño years in recent decades.

However, trends in annual total rainfall post-1900 reveal modest increases over the MDB catchment (Collins and Della-Marta, 2002) (Fig. 4A). This may indicate that both periods simply reflect natural interdecadal rainfall variability. But strong global warming only became evident in the last few decades of the twentieth century, following a cooling trend (mainly in the Northern Hemisphere) in the 1940s and 1950s, so it is arguable that the more recent Australian rainfall trends may be related to the enhanced greenhouse effect. Comparison with projected rainfall changes due to the enhanced greenhouse effect (see below) do not, however, show strong agreement, so perhaps the enhanced greenhouse signal in Australian rainfall has yet to dominate over natural variability.

The annual total rainfall averaged over all of the MDB for each year since 1950 (Fig. 4B) has decreased by 11.0 mm/decade, dominated by high year-to-year variability. However, since 1900 the annual mean MDB rainfall actually shows an increase of 6.8 mm/decade (Fig. 4A) and further highlights the dominance of interannual and interdecadal rainfall variability, so that linear trends are highly dependent on the period of analysis. Year-to-year changes in mean rainfall are generally associated with similar changes in extreme rainfall: during high rainfall periods, there are a larger number of rain days, days with significant rainfall (>10 mm) or days with extreme rainfall (daily totals above the 1961–1990 mean 95th percentile level). However, on average, there have been decreases in the frequency of each of these measures over the catchment since 1950. Similar decreases are evident in the intensity of extreme rainfall.

VI. CLIMATE CHANGE PROJECTIONS

Regional climate change scenarios, that is, projections of possible or plausible climate changes as a result of the enhanced greenhouse effect, were first mooted in the late 1970s, and the first for the Australian region in the early 1980s (Pittock and Salinger, 1982). These were based on primitive general circulation models of the climate, with grossly simplified oceans, crude analogies with past warm epochs and ensembles of individual warm versus cool years, and elementary theoretical reasoning. Despite the inadequate basis of these scenarios, they suggested that the Australian monsoon might increase in strength and the westerly rain-bearing systems in the south might move further south, conclusions that remain relevant today. Pittock (1983) went

A

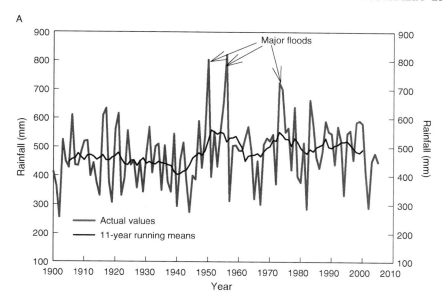

B

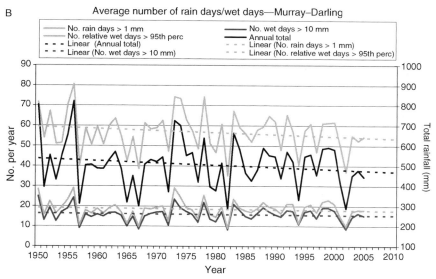

Figure 4 (A) Annual (blue line) and 11-year-running mean rainfall averaged over the Murray–Darling Basin region from 1900 to 2005. Source: Bureau of Meteorology. (B) Time series of MDB annual mean rainfall, frequency of wet days, frequency of days with significant rainfall (>10 mm), and frequency of days with extreme rainfall (daily totals above the 1961–1990 mean 95th percentile level) from 1950 to 2005 (courtesy of Dean Collins, Bureau of Meteorology). (See Color Plate 3.)

further and suggested that the rainfall changes around 1945–1946 might prove to be partly due to global warming and a good analogy of what might be expected in future.

The first specific analysis of possible changes in the MDB, with implications for irrigated agriculture, was published by Pigram *et al.* (1992). This was based on climate change scenarios issued by CSIRO in 1992, but did not discuss floods.

Several later studies were undertaken on water resources including parts of the MDB. These include Chiew *et al.* (1995) and Schreider *et al.* (1996, 1997) and were based on the CSIRO (1992, 1996) climate change scenarios. The 1996 scenarios were based on both simple mixed layer and more realistic coupled ocean-atmosphere climate models and had a wide range of uncertainty in rainfall changes.

More recent studies used the CSIRO (2001) scenarios. These scenarios were based on simulations from nine coupled ocean-atmosphere climate models and using the full range of projected global warming as given in the Intergovernmental Panel on Climate Change report in 2001 (IPCC, 2001) in combination with projected regional changes obtained from nine climate models. The ranges of change allow for uncertainty in human behavior (e.g., future emissions of greenhouse gases), as well as climate science uncertainty (e.g., differences in the response of climate models).

However, the most recent and so far unpublished analyses have used 23 global and regional climate model simulations to see how well these models capture features of the present climate over southeastern Australia. Statistical methods were employed to test whether the simulations satisfactorily resemble observed patterns of mean sea level pressure, temperature, and precipitation. On the basis of the results of statistical analyses, the nine better performing models were selected to construct temperature and rainfall change projections for the selected domain. However, only seven models, which have point potential evaporation data, have been selected to construct future projections for potential evaporation.

Climate change projections for the MDB are presented here for the years 2030 and 2070 for different emission scenarios. These scenarios are the IPCC special report on emission scenarios (SRES) without policies to reduce greenhouse gas emissions, the IPCC scenario for stabilizing CO_2 concentrations at 550 ppm by the year 2150 and the IPCC scenario for stabilizing CO_2 concentrations at 450 ppm by the year 2100. Results from the global climate model (GCM) simulations were expressed as a change per degree of global warming, then scaled for 2030 and 2070 using the IPCC emission scenarios. The range in each projection represents the range in the results of the nine GCMs.

Figure 5 indicates a general increase in temperature. The inland regions show a warming of between $+0.5$ and $+2.0\,°C$ by 2030 and between $+1.0$ and

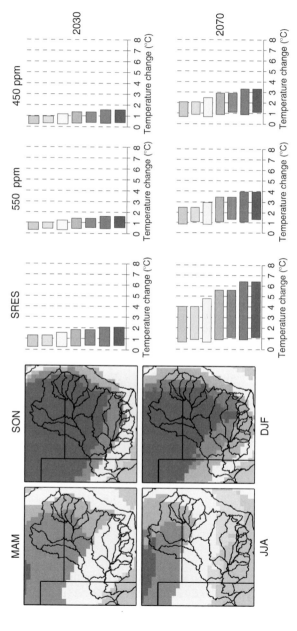

Figure 5 Projected temperature changes for the Murray–Darling Basin by 2030 and 2070. Units are in degree Celsius. (See Color Plate 4.)

+6.5 °C by 2070 during spring and summer for SRES emission scenarios. The southern part of the basin shows less warming. Relatively smaller increases are projected for stabilized scenarios for 550 and 450 ppm. During autumn and winter increases are between +0.5 and +2.0 °C by 2030 and between +1.0 and +6.5 °C by 2070, except for a small region in the inland which shows greater warming in summer. Southern coastal regions show relatively less warming.

Rainfall projections are summarized in Fig. 6. Both winter and spring show predominant decreases, in which spring shows a stronger decrease. Projected rainfall changes during winter and spring are between +5% and –20% by 2030 and +15% and –60% by 2070 for the SRES emission scenarios. Relatively smaller changes are indicated for stabilized emission scenarios. Summer changes in rainfall as large as 40% increases are indicated in the western and central parts of the Basin by 2070, with similar decreases in the south. In autumn increases are more general, except in the northeast. Projected high case changes are reduced by 40% in the 450 ppm case.

Point potential evaporation increases between 5% and 15% by 2030 and between 15% and 45% by 2070 for SRES scenarios are shown in Fig. 7. The ranges are relatively smaller for stabilized scenarios for 550 and 450 ppm.

Studies of flood occurrence and impacts require information about extreme rainfall intensity and frequency for event durations ranging from hours, in the case of flash floods, to multiple days for large area flooding. Flood impact models also rely on information about how rapidly the average rainfall intensity increases with decreasing area, that is, depth–area curves. These relationships are likely to be altered by climate change. Quantifying these changes ideally requires very fine spatial resolution (of the order of 4 km) and high temporal resolution (hourly) using both dynamical modeling and statistical methods. This is extremely computationally intensive to do Australia-wide, and to date computations have been done at coarser spatial (65 km) and time (daily) resolutions.

While some regions show a tendency toward drier seasonal-average conditions under enhanced greenhouse conditions, it does not necessarily follow that extreme daily rainfall events will become less frequent or severe in those seasons. Previous studies based on daily rainfall data from various climate models have indicated marked increases in the intensity and frequency of extreme daily rainfall events under enhanced greenhouse conditions for the Australian region (Whetton et al., 1993; Suppiah, 1994; Fowler and Hennessy, 1995; Hennessy et al., 1997; Whetton et al., 2002).

Suppiah (1994) used the coarse resolution (5.6° × 3.2°) CSIRO9 climate model to examine the synoptic situations leading to heavy rain events in central Australia. He found these were associated with more intense cutoff lows and that under doubled CO_2 conditions the lows were more intense and produced heavier rainfall. This extended to the southeast into the MDB catchment. Such a study needs to be repeated at higher spatial resolution.

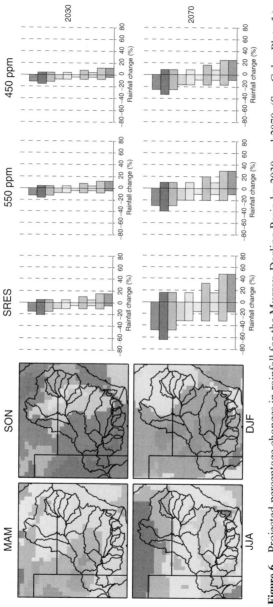

Figure 6 Projected percentage changes in rainfall for the Murray–Darling Basin by 2030 and 2070. (See Color Plate 5.)

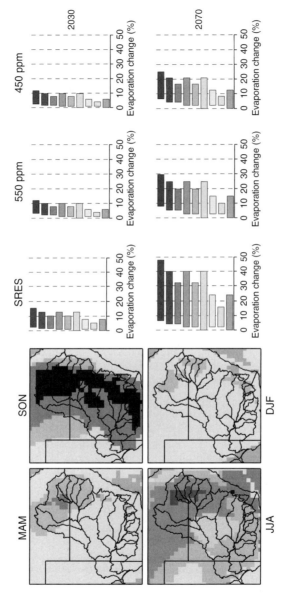

Figure 7 Projected percentage changes in point potential evaporation for the Murray–Darling Basin by 2030 and 2070. (See Color Plate 6.)

The requirement of daily rainfall data limits our ability to capture the full range of uncertainty inherent in climate change science since we only have access to daily fields from the Australian climate models. In addition, the coarse resolution of climate models means that they are unable to represent either the full intensity of extreme rainfall events or the strong spatial variability of these events. Hennessy *et al.* (2004) present the most recent projections of extreme rainfall changes for NSW. These methods have been extended to all of southeastern Australia and are based on daily outputs from five Australian climate model simulations spatially interpolated to a common 0.5° grid.

Projections may be used to convey information about the possible direction of change in extreme rainfall intensity or the likelihood of a particular change occurring due to consistency between the climate models used. Projections of likely regions of increase or decrease in annual and seasonal extreme 3-day rainfall events (as intermodel consistencies between the five models) are presented in Fig. 8 for southeastern Australia for the climate of 2070, by which time the climate change signal may dominate over the large natural variability.

These projections show the intermodel consistency for the direction of change of the 10 highest rainfall events from each of the 5 models by 2070. The climate change signal is mixed for annual extreme rainfall events, but most models indicate that extremes will increase in intensity in southwestern NSW and decrease in intensity further north. In most other regions, there is less intermodel consistency in the direction of change. In summer, autumn, and winter, there is greater consistency as to the likely direction of change for 3-day extreme rainfall events. During summer, most models indicate decreases in rainfall extremes in the central western portion of the MDB catchment and increases in the northern and southern portions. Autumn and winter extreme rainfalls are projected to increase by most models for the MDB, especially in the central parts of the catchment. In spring, decreases in the intensity of extreme rainfall events are projected for the northern half of the catchment, with increases apparent in the southern half of the catchment. Similar patterns of change (not shown) are seen for the 1-day events.

Caution must be used in interpreting these results as in general rainfall extremes are found to increase in intensity when the total rainfall increases (Fig. 4B). This does not seem to be fully borne out between Figs. 6 and 8, but this may be because in Fig. 6 results are from nine climate models of which only one is Australian, whereas in Fig. 8 results are from five climate models, all of which are Australian. Further work is needed to develop a more consistent picture allowing greater confidence in projections.

Changes in rainfall intensity, which are critical for projections of flood behavior, are probably not fully captured in the current analyses with the global and regional modeling, although the preceding results provide a first approximation and do suggest that significant changes are likely to occur. Changes in the frequency and intensity of tropical cutoff lows affecting the

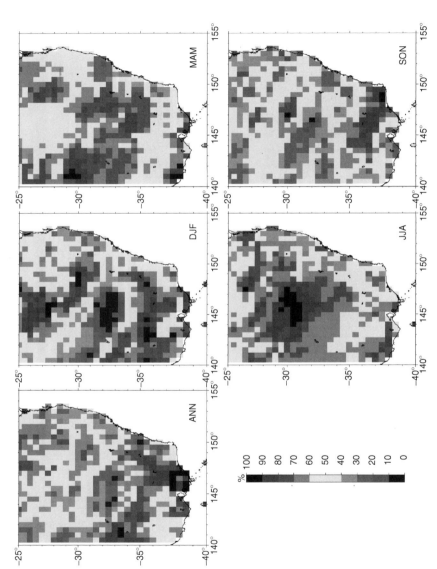

Figure 8 The likelihood (%) of an increase in 3-day extreme rainfall for 2070 relative to the current climate. Red regions denote regions where most of the 5 models simulate a decrease in the intensity of the 10 highest rainfall events and blue regions denote those where most models project an increase in intensity of heavy rainfall events. ANN, annual; DJF, summer; MAM, autumn; JJA, winter; SON, spring. (See Color Plate 7.)

northern section of the MDB, especially those resulting from former tropical cyclones that move inland, are uncertain, as are changes in cutoff lows in winter and spring in the south, and also changes in thunderstorm intensities as these small-scale events are poorly represented in the models.

Future trends in tropical cyclone numbers and tracks are unclear. The main difficulty is that despite many years of work, it is still uncertain how the ENSO will change in a warmer world, with different climate model experiments suggesting different changes (IPCC, 2001, especially Working Group I, and Cai and Whetton, 2000 for an Australian analysis). Since ENSO has a strong effect on tropical cyclone numbers in the Australian region, and also regarding the strength of the summer monsoon, this creates considerable uncertainty about future predictions.

There is more convincing evidence of future increases in cyclone intensities (IPCC, 2001). The main physical argument is that with projected increases in sea surface temperatures, more energy is available for tropical cyclones and thus maximum storm intensities increase. Both theoretical techniques and model simulations support this view. However, climate models of tropical cyclones until very recently have been run with horizontal resolutions no finer than 18 km, which is relatively coarse compared with the actual processes taking place in tropical cyclones. Research elsewhere suggest that resolutions finer than 5 km are needed to fully capture the important processes for intensity changes in tropical cyclones.

Recently, the CSIRO Conformal Cubic Atmospheric Model (CCAM) has been used to downscale the outputs from the CSIRO GCMs to provide high-quality climate projections over the Australian region. These outputs are available at a grid spacing of approximately 65 km and they have been examined to identify tropical cyclone like vortices (TCLVs) objectively using software that searches for closed isobars and refines these searches through selection criteria that require specification of a threshold wind speed, a warm core, and other parameters.

The results from this analysis are shown in Fig. 9 for the model climate corresponding to the current period (1960–2000). These results are compared with the observations from the Bureau of Meteorology tropical cyclone database for the period 1970–2000. The results show a good qualitative agreement in the preferred regions of occurrence under present climatic conditions with maxima in the Coral Sea, Gulf of Carpentaria, and off the coast of north Western Australia. However, in all regions the modeled frequencies are less than observed.

A similar analysis of TCLVs for the 2030 (2010–2050) and 2070 (2050–2090) climates shows a decrease in the total number of storm occurrences by approximately 13.5% for 2030 and 30.5% in 2070. However, it is not known at this stage if this decrease in frequency is peculiar to the CCAM results only. Other climate models should be used to investigate this change further. This decrease in cyclone numbers is accompanied by an increase in intensity (decrease in central pressure and increase in maximum wind speeds) in a

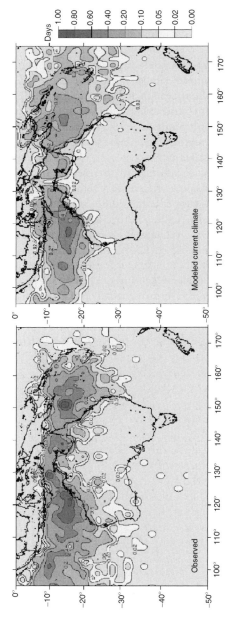

Figure 9 Comparison of observed (1960–2000) tropical cyclone numbers for the Australian region with those simulated by the regional climate model CCAM. Shown are the number of tropical cyclone days per year in each 2 × 2 grid box. (See Color Plate 8.)

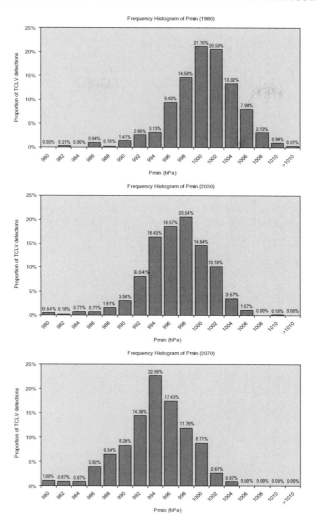

Figure 10 Frequency distribution of minimum central pressure of Australian region tropical cyclones for the current climate (top), the climate of 2030 (middle), and that of 2070 (bottom).

warmer world (Fig. 10). This implies a significant increase in rainfall intensity associated with tropical cyclones and the tropical lows which result from such cyclones as they move inland.

However, with a grid spacing of 65 km, it is impossible to simulate the true intensity of the cyclones, and high-resolution modeling techniques must be used if changes in the intensity of the storms are to be quantified. For example, in parallel simulations of TC Bobby, a simulation with a grid spacing of 15 km produced a minimum central pressure of 955 hPa and a maximum wind speed of 55 m/s. When the grid spacing was reduced to 5 km, the simulated intensity of the cyclone was increased to a minimum central pressure of 933 hPa (935 hPa was estimated) and a maximum wind speed of 75 m/s.

Recent claims suggest that the number of category 4 and 5 tropical cyclones (the most intense) have increased by tens of percent in number over recent decades in several ocean basins (Emanuel, 2005a,b; Webster *et al.*, 2005), whereas climate modeling results suggest much smaller increases in intensity (IPCC, 2001; Knutsen and Tuleya, 2004). Hoyas *et al.* (2006) found that the claimed increase in intense cyclones in several ocean basins is correlated best with increasing sea surface temperatures in each basin. It has been argued that these results may merely reflect dubious observational data (Pielke, 2005; Pielke *et al.*, 2005), but it is perhaps more likely that the disagreement between observations and models may reflect shortcomings in the model simulations, perhaps due to inadequate spatial resolution or other simplifications in the models.

What does seem certain is that rainfall intensities associated with former tropical cyclones are likely to be more intense (Knutsen and Tuleya, 2004). This may also apply to cutoff lows (Suppiah, 1994) and even to individual thunderstorms. Where these occur in regions of generally increased summer and autumn rainfall, more intense runoff events seem very likely.

VII. CONCLUSIONS

Floods in the MDB can be due to local severe storms leading to flash flooding, or to heavy rains from much larger-scale systems leading to basin-wide great floods which can take months to travel down stream. The Basin is subject to large-scale rain events from diverse synoptic origins, mainly tropical lows in the summer half year in the northern region and fronts and cutoff lows in the southern sector during the winter half year. Great year-to-year variability is associated with fluctuations in the ENSO regime, and in the past there have been marked interdecadal variations in rainfall, some of which have occurred abruptly.

Changes due to the enhanced greenhouse effect will include higher temperatures and increased potential evaporation, and likely increases in rainfall intensity associated with major storm systems, including cutoff lows of both tropical and midlatitude origin. Annual mean rainfall is generally expected to increase in northern regions of the MDB, especially inland, but to decrease in the southernmost areas south of the Murray River.

However, the nature of changes to ENSO are still unclear, as is the extent of increases in the intensity of tropical cyclones and other tropical and midlatitude low-pressure systems. The frequency and preferred paths of tropical cyclones and cutoff lows may also change. The influence of lows related to cold fronts and cutoff lows associated with the midlatitude wester-lies may decrease, leading to reduced rainfall and flooding in the southern-most regions of the Basin, especially in the winter half year. Changes in rainfall intensity in models are scale-dependent, with finer resolution models generally predicting larger increases. More work is needed to resolve these uncertainties, and to apply flood routing to rainfall scenarios.

Taking all these uncertainties into account, any accurate quantitative estimate of changes in flood frequency in the MDB due to the enhanced greenhouse effect is at this stage premature. On balance, it seems likely that extreme rainfall events may become more extreme, especially in the northern areas in the December–May half year. If this is combined with a general degradation in vegetative cover (Raupach et al., 2005) leading to faster flood runoff, more great floods may be expected, although interspersed with periods of increasingly severe drought (Nicholls, 2004). This may mobilize more sediment, increase bank erosion, and change river morpho-logy, with large sand plugs developing. On the other hand, widespread reforestation in the upper catchments, perhaps to store carbon, could lead to reductions in runoff (Herron et al., 2002; Jones et al., 2002).

Climate change is now a "hot topic" with new research results appearing almost weekly on issues such as possible changes to tropical cyclones and ENSO and other factors affecting rates of change such as biospheric feedbacks. Readers are therefore advised to keep up to date with the emerging literature.

ACKNOWLEDGMENTS

Thanks are due to fellow members of the Climate Impacts, Adaptation and Risk group in CSIRO Marine and Atmospheric Research, and in particular to Tony Rafter for help with graphics and Ian Smith for valuable comments. Thanks are also due to Dean Collins and David Jones of the Australian Bureau of Meteorology in Melbourne for data and Figures, Martin Thoms of the University of Canberra for help with Table 1, and Mike Smith of the National Museum of Australia for additional comments.

REFERENCES

Abbs, D., Aryal, S., Campbell, E., McGregor, J., Nguyen, K., Palmer, M., Rafter, T., Watterson, I. and Bates, B. (2005) "Projections of Extreme Rainfall and Cyclones." Final Report to the Australian Greenhouse Office. CSIRO, Aspendale.

Annette, P. (1978) Some aspects of troughs in the easterlies over Southern Hemisphere continents. *Meteorological Note 100*, Bureau of Meteorology, Melbourne.

Bowler, J.M. (1998) Willandra lakes revisited: Environmental framework for human occupation. *Archaeol. Oceania* **33**, 120–155.

BTE (2001) *Economic Costs of Natural Disasters in Australia.* Bureau of Transport Economics, Canberra.

Cai, W. and Whetton, P.H. (2000) Evidence for a time-varying pattern of greenhouse warming in the Pacific Ocean. *Geophys. Res. Lett.* **27**, 2577–2580.

Cai, W., Whetton, P.H. and Pittock, A.B. (2001) Fluctuations in the relationship between ENSO and northeast Australian rainfall. *Clim. Dyn.* **17**, 421–432.

Cai, W., Whetton, P.H. and Karoly, D.J. (2003) The response of the Antarctic Oscillation to increasing and stabilized atmospheric CO_2. *J. Clim.* **16**, 1525–1538.

Chiew, F.H.S., Whetton, P.H., McMahon, T.A. and Pittock, A.B. (1995) Simulations of the impact of climate change on runoff and soil moisture in Australian catchments. *J. Hydrol.* **167**, 121–147.

Collins, D.A. and Della-Marta, P.M. (2002) "Atmospheric Indicators for the State of the Environment Report 2001." Technical Report No. 74 Bureau of Meteorology, Australia, 25p.

Craik, W. (2005) "Weather, Climate, Water and Sustainable Development." World Meteorological Day Address 2005, Bureau of Meteorology, Melbourne.

CSIRO (1992) "Climate Change Scenarios for Australia." CSIRO Division of Atmospheric Research, Aspendale.

CSIRO (1996) "Climate Change Scenarios for Australia." CSIRO Division of Atmospheric Research, Aspendale.

CSIRO (2001) "Climate Change Scenarios for Australia." CSIRO Division of Atmospheric Research, Aspendale.

Deacon, E.L. (1953) Climatic changes in Australia since 1880. *Aust. J. Phys.* **6**, 209–218.

Devin, L.B. and Purcell, D.L. (1983) "Flooding in Australia." Water 2000: Consultants Report 11, AGPS, Canberra.

Emanuel, K. (2005a) Increasing destructiveness of tropical cyclones over the past 30 years. *Nature* **436**, 686–688.

Emanuel, K. (2005b) Emanuel replies. *Nature* **438**, December 22/29, DOI: 10.1038/nature.04427.

Fowler, A.M. and Hennessy, K.J. (1995) Potential impacts of global warming on the frequency and magnitude of heavy precipitation. *Nat. Hazards* **11**, 283–303.

Franks, S.W. (2002) Identification of a change in climate state using regional flood data. *Hydrol. Earth Syst. Sci.* **6**, 11–16.

Franks, S.W. and Kuczera, G. (2002) Flood frequency analysis: Evidence and implications of secular climate variability, New South Wales. *Water Resour. Res.* **38**, art. no. 1062.

Gentilli, J. (1971) Climatic fluctuation: Climates of Australia and New Zealand. *World Surv. Climatol.* **13**, 89.

Grimmer, L.C. (1988) The rain event of 25–26 January 1984 over the Australian Capital Territory and surrounding districts of New South Wales. *Meteorological Note 180*, Bureau of Meteorology, Melbourne.

Hennessy, K.J., Gregory, J.M. and Mitchell, J.F.B. (1997) Changes in daily precipitation under enhanced greenhouse conditions. *Clim. Dyn.* **13**, 667–680.

Hennessy, K.J., Suppiah, R. and Page, C.M. (1999) Australian rainfall changes, 1910–1995. *Aust. Meteorol. Mag.* **48**, 1–13.

Hennessy, K.J., McInnes, K., Abbs, D., Jones, R., Bathols, J., Suppiah, R., Ricketts, J., Rafter, T., Collins, D. and Jones, D. (2004) "Climate Change in New South Wales. Part 2: Projected Changes in Climate Extremes." CSIRO Atmospheric Research.

Herron, N., Davis, R. and Jones, R. (2002) The effects of large-scale afforestation and climate change on water allocation in the Macquarie River catchment, NSW, Australia. *J. Environ. Manag.* **65**, 369–381.

Hesse, P.P., Magee, J.W. and van der Kaars, S. (2004) Late Quaternary climates of the Australian arid zone: A review. *Quat. Int.* **118–119**, 87–102.

Holland, G.J., Lynch, A.H. and Leslie, L.M. (1987) Australian east-coast cyclones, I, Synoptic overview and case study. *Mon. Weather Rev.* **115**, 3024–3036.

Hope, G., Kershaw, A.P., van der Kaars, S., Xiangjun, S., Liew, P.-M., Heusser, L.E., Takahara, H., McGlone, M., Miyoshi, N. and Moss, P.T. (2004) History of vegetation and habitat change in the Austral-Asian region. *Quat. Int.* **118–119**, 103–126.

Hoyas, C.D., Agudelo, P.A., Webster, P.J. and Curry, J.A. (2006) Deconvolution of the factors contributing to the increase in global hurricane intensity. *Science* **312**, 94–97.

Hulme, M., Barrow, E.M., Arnett, N.W., Harrison, P.A., Johns, T.C. and Downing, T.E. (1999) Relative impacts of human-induced climate change and natural climate variability. *Nature* **397**, 688–691.

IPCC (2001) "Climate Change 2001: Synthesis Report, Contribution of Working Groups I, II and III to the IPCC Third Assessment Report." Cambridge University Press, Cambridge. See www.ipcc.ch for this and more detailed reports.

Jones, R., Whetton, P., Walsh, K. and Page, C. (2002) "Future Impacts of Climate Variability, Climate Change and Land Use Change on Water Resources in the Murray Darling Basin." CSIRO, Aspendale.

Jones, R.N. and Pittock, A.B. (2002) Climate change and water resources in an arid continent: Managing uncertainty and risk in Australia. In: *Climatic Change: Implications for the Hydrological Cycle and for Water Management, Advances in Global Change Research* (Ed. by M. Beniston), Vol. 10, pp. 465–501. Kluwer Academic Publishers, Dordrecht, Boston.

Kingsford, R.T. (2000) Ecological impacts of dams, water diversions and river management on floodplain wetlands in Australia. *Aust. Ecol.* **25**, 109–127.

Kirkup, H., Brierley, G., Brooks, A. and Pitman, A. (1998) Temporal variability of climate in south-eastern Australia: A reassessment of flood- and drought-dominated regimes. *Aust. Geogr.* **29**, 241–255.

Knutsen, T.R. and Tuleya, R.E. (2004) Impact of CO_2-induced warming on simulated hurricane intensity and precipitation: Sensitivity to the choice of climate model and convective parameterization. *J. Clim.* **17**, 3477–3495.

Kraus, E.B. (1954) Secular changes of tropical rainfall regime of southeast Australia. *Q. J. R. Meteorol. Soc.* **80**, 591–601.

Kraus, E.B. (1955) Secular changes of east-coast rainfall regimes. *Q. J. R. Meteorol. Soc.* **81**, 430–439.

Kraus, E.B. (1963) Recent changes of east-coast rainfall regimes. *Q. J. R. Meteorol. Soc.* **89**, 145–146.

Lavery, B., Kariko, A.B. and Nicholls, N. (1992) A high-quality historical rainfall data set for Australia. *Aust. Meteorol. Mag.* **40**, 33–39.

Lavery, B., Joung, G. and Nicholls, N. (1997) An extended high-quality historical rainfall data set for Australia. *Aust. Meteorol. Mag.* **46**, 27–38.

Marshall, G.J. (2003) Trends in the Southern Annular Mode from observations and reanalysis. *J. Clim.* **16**, 4134–4143.

Meneghini, B., Simmonds, I. and Smith, I. (2006) Association between Australian rainfall and the Southern Annular Mode. *Int. J. Climatol.* (in press).

Miller, G.H., Fogel, M.L., Magee, J.W., Gagan, M.K., Clarke, S.J. and Johnson, B.J. (2005) Ecosystem collapse in Pleistocene Australia and a human role in megafaunal extinction. *Science* **309**, 287–290.

Moore, R.J. (1988) The drought relieving rains of 20–27 May 1981. *Meteorological Note 182*, Bureau of Meteorology, Melbourne.

Nicholls, N. (2000) The insignificance of significance testing. *Bull. Am. Meteorol. Soc.* **81**, 981–986.

Nicholls, N. (2004) The changing nature of Australian droughts. *Clim. Change* **63**, 323–336.

Nicholls, N. and Lavery, B. (1992) Australian rainfall trends during the twentieth century. *Int. J. Climatol.* **12**, 153–163.

Nicholls, N., Lavery, B., Fredericksen, C., Drosdowsky, W. and Torok, S. (1996) Recent apparent changes in relationships between the El Niño-Southern Oscillation and Australian rainfall and temperature. *Geophys. Res. Lett.* **23**, 3357–3360.

Pielke, R.A., Jr. (2005) Are there trends in hurricane destruction? *Nature* **438**, December 22/29, DOI: 10.1038/nature.04426.

Pielke, R.A., Jr., Landsea, C., Mayfield, M., Laver, J. and Pasch, R. (2005) Hurricanes and global warming. *Bull. Am. Meteorol. Soc.* **86**, 1571–1575.

Pigram, J.J., Shaw, K.L. and Coelli, M.L. (1992) "Climate Change and Irrigated Agriculture." Centre for Water Policy Research, Armidale, NSW. Position Paper for the Greenhouse Information Program, Department of Arts, Sport, the Environment and Territories, April 1992.

Pittock, A.B. (1975) Climatic change and the pattern of variation in Australian rainfall. *Search* **6**, 498–504.

Pittock, A.B. (1983) Recent climatic change in Australia: Implications for a CO_2-warmed Earth. *Clim. Change* **5**, 321–340.

Pittock, A.B. and Salinger, M.J. (1982) Toward regional scenarios for a CO_2-warmed Earth. *Clim. Change* **4**, 23–40.

Raupach, M., Briggs, P., King, E., Schmidt, M., Paget, M., Lovell, J. and Canadell, P. (2005) "Impacts of Decadal Climate Trends on Australian Vegetation." EOC Symposium, Canberra, February 15.

Schreider, S.Y., Jakeman, A.J., Pittock, A.B. and Whetton, P.H. (1996) Estimation of possible climate change impacts on water availability, extreme flow events and soil moisture in the Goulburn and Ovens basins, Victoria. *Clim. Change* **34**, 513–546.

Schreider, S.Y., Jakeman, A.J., Whetton, P.H. and Pittock, A.B. (1997) Estimation of climate impacts on water availability and extreme flow events for snow-free and snow-affected catchments of the Murray-Darling basin. *Aust. J. Water Resour.* **2**, 35–46.

Shulmeister, J., Goodwin, I., Renwick, J., Harle, K., Armand, L., McGlone, M.S., Cook, E., Dodson, J., Hesse, P.P., Mayewski, P. and Curran, M. (2004) The Southern Hemisphere westerlies in the Australian sector over the last glacial cycle: A synthesis. *Quat. Int.* **118–119**, 23–53.

Singh, G. and Geissler, E.A. (1985) Late Caenozoic vegetation history of vegetation, fire, lake levels and climate at Lake George, New South Wales, Australia. *Philos. Trans. R. Soc. London B* **311**, 379–447.

Smith, I.N. (2004) An assessment of recent trends in Australian rainfall. *Aust. Meteorol. Mag.* **53**, 163–173.

Smith, M. (2005) Moving into the southern deserts: An archaeology of dispersal and colonisation. In: *23°S Archaeology and Environmental History of the Southern Deserts* (Ed. by M. Smith and P. Hesse), pp. 92–107. National Museum of Australia Press, Canberra.

Srikanthan, R. and McMahon, T.A. (2001) Stochastic generation of annual, monthly and daily climate data: A review. *Hydrol. Earth Syst. Sci.* **5**, 653–670.

Suppiah, R. (1994) Synoptic aspects of wet and dry conditions in central Australia: Observations and GCM simulations for $1 \times CO_2$ and $2 \times CO_2$ conditions. *Clim. Dyn.* **10**, 395–405.

Suppiah, R. (2004) Trends in the Southern Oscillation phenomenon and Australian rainfall and changes in their relationship. *Int. J. Climatol.* **24**, 269–290.

Suppiah, R. and Hennessy, K.J. (1998) Trends in total rainfall, heavy rain events and number of dry days in Australia, 1910–1990. *Int. J. Climatol.* **18**, 1141–1164.

Tapper, N. and Hurry, L. (1993) *Australia's Weather Patterns: An Introductory Guide.* Dellasta, Mount Waverley, Victoria.

Thompson, D.W.J. and Wallace, J.M. (2000) Annular modes in the extratropical circulation, Part I: Month-to-month variability. *J. Clim.* **13**, 1000–1016.

Thompson, D.W.J., Wallace, J.M. and Hegerl, G.C. (2000) Annular modes in the extratropical circulation, Part II: Trends. *J. Clim.* **13**, 1018–1036.

Thoms, M., Sheldon, F. and Crabb, P. (2004) A Hydrological Perspective on the Darling River. In: *The Darling* (Ed. by R. Breckwoldt, R. Boden and J. Andrew), pp. 332–347. Murray-Darling Basin Commission, Canberra.

Trenberth, K.L. (1976) Fluctuations and trends in indices of the Southern Hemisphere circulation. *Q. J. R. Meteorol. Soc.* **102**, 65–75.

Troup, A.J. (1965) The "Southern Oscillation." *Q. J. R. Meteorol. Soc.* **91**, 490–506.

Vives, B. and Jones, R. (2005) "Detection of Abrupt Changes in Australian Decadal Rainfall Variability (1890–1989)." CSIRO Technical Paper no.73. CSIRO Marine and Atmospheric Research. (Electronic copy only.)

Warner, R.F. (1987) The impacts of alternating flood- and drought-dominated regimes on channel morphology at Penrith, New South Wales, Australia. *Int. Assoc. Hydrol. Sci.* **168**, 327–338.

Warner, R.F. (1995) Predicting and managing channel change in southeast Australia. *Catena* **25**, 403–418.

Webster, P.J., Holland, G.J., Curry, J.A. and Chang, H.-R. (2005) Changes in tropical cyclone number, duration, and intensity in a warming environment. *Science* **309**, 1844–1846.

Whetton, P.H., Adamson, D. and Williams, M. (1990) Rainfall and river flow variability in Africa, Australia, East Asia linked to El Niño–Southern Oscillation events. *Geol. Soc. Aust. Symp. Proc.* **1**, 71–82.

Whetton, P.H., Fowler, A.M., Haylock, M.R. and Pittock, A.B. (1993) Implications for floods and droughts in Australia of climate change due to the enhanced greenhouse effect. *Clim. Change* **25**, 289–317.

Whetton, P.H., Suppiah, R., McInnes, K.L., Hennessy, K.J. and Jones, R.N. (2002) *Climate Change in Victoria: High Resolution Regional Assessment of Climate Change Impacts.* Victorian Department of Natural Resources and Environment, Melbourne, 44p.

Wright, W.J. (1988a) The low latitude influence on winter rainfall in Victoria, southeastern Australia. I. Climatological aspects. *Int. J. Climatol.* **8**, 437–462.

Wright, W.J. (1988b) The low latitude influence on winter rainfall in Victoria, southeastern Australia. II. Relationships with the southern oscillation and Australian region circulation. *Int. J. Climatol.* **8**, 547–576.

Wright, W.J. (1997) Tropical-extratropical cloudbands and Australian rainfall: I. Climatology. *Int. J. Climatol.* **17**, 807–829.

Floods Down Rivers: From Damaging to Replenishing Forces

SAM LAKE, NICK BOND AND PAUL REICH

SUMMARY

Floods are major forces acting on physical and biological processes within rivers, but vary substantially in their effects according to location within a river network. In upland streams, the rapid and dramatic physical disturbance caused by floods often results in the decline of local biota, however faunal resilience is high and recovery from floods is generally fast. In contrast, floods in lowland rivers act as a replenishing force, connecting floodplain habitat and stimulating high rates of production and growth. In these systems, adaptations are geared more toward capitalizing on floods than on avoidance and recovery. Despite these differences, floods play a key role in the dynamics and long-term persistence of aquatic ecosystems in both settings.

I. INTRODUCTION

Floods, whether they are rapid events in small streams or prolonged events in large floodplain rivers, are clearly a major force shaping both the

ADVANCES IN ECOLOGICAL RESEARCH VOL. 39
0065-2504/06 $35.00
DOI: 10.1016/S0065-2504(06)39003-4

physicochemical conditions and the biology of streams and rivers. Their opposite disturbance, namely droughts, also exert a powerful influence on the ecology of flowing waters, but at present their effects are far less understood than those generated by floods.

There is already a large body of literature dealing with the effects of floods on a wide variety of aquatic ecosystems, from intermittent systems, such as flood runners and streams in arid regions, to perennial systems—from cold, turbulent montane streams to large, turbid floodplain rivers. A number of conceptual models, such as patch-dynamics concept (Pickett and White, 1985) and the flood pulse concept (Junk et al., 1989), describe the effects of floods in upland and lowland streams, and others, such as the river continuum concept (RCC) (Vannote et al., 1980), explore longitudinal changes in aquatic ecosystems. However, despite the existence of these models, the contrasting role that floods play in constrained steep gradient upland streams and low-gradient unconfined lowland systems has rarely been considered in concert. Thus, a major aim of this chapter is to draw attention to the similarities and differences in the physical characteristics of floods and their ecological effects in upland and lowland ecosystems. As well as covering recent developments in the ecological understanding of floods since the review of Lake (2000), we highlight the damaging and replenishing nature of floods in different ecosystems drawing, where appropriate, on specific examples. We also place some emphasis on examples drawn from the Murray–Darling Basin, a system that harbors a wide variety of watercourses and has been a major focus for aquatic ecosystem research in Australia. We begin by describing the physical nature of floods in upland and lowland systems before examining the ecological response and importance of floods in each of these settings.

II. PERTURBATION: DISTURBANCE AND RESPONSE

It is important to view an event such as a flood as a perturbation comprising two distinct sequential parts. The flood event itself is a disturbance—a damaging event that has the *potential* to damage habitat and resources used by organisms (Pickett and White, 1985; Lake, 2000). It is the potential floods have to cause damage that necessitates a definition based on their physical strength rather than on any biological response (Rykiel, 1985; Lake, 1990; Poff, 1992) in order to make objective comparisons between flood events. Such physical parameters include flow rate, intensity (shape of hydrograph), average recurrence interval, spatial extent, duration, and predictability (Smith and Ward, 1998; Gordon et al., 2004).

During and following the disturbance, there is the response by the biota (Lake, 2000). The response is seen as made up of two properties: resistance or the capacity to withstand the disturbance, and resilience or the capacity to recover from the disturbance (Webster *et al.*, 1983). Responses to disturbance vary tremendously between types of flowing waters and may also vary within stream types, even to the same type and magnitude of disturbance (Poff, 1992; Matthaei *et al.*, 2004).

III. THE PHYSICAL DIMENSION: FLOODS AS DISTURBANCES

Floods are often defined as occurring when water leaves a river channel and flows out over the banks onto the surrounding land (Ward, 1978). This is the normal definition that people accept as such floods can inflict considerable economic damage and are associated with floodplain inundation. In upland and constrained reaches of rivers, however, floods may rarely leave the confines of the channel, the increased discharge being accommodated instead by dramatic increases in the velocity of the water. Floods are pulse disturbances—that is the event is clearly defined as a pulse—with a hydrograph consisting of a rather steep rising limb to reach a peak and then a gentler receding limb back to normal flow. In general, the physical nature of floods vary predictably as one moves down through a catchment from being short-term, high-energy events in upland streams to prolonged large-scale events on floodplains, which in the latter case may inundate the surrounding floodplain for weeks or months on end (Fig. 1). For example, dryland rivers of inland Australia, such as the Darling, Paroo, and Cooper Creek, are remarkable for their high flow variability (McMahon *et al.*, 1992; Puckridge *et al.*, 1998), and when the occasional floods do occur, the extensive low-lying floodplain means they are of long duration as they move slowly across the landscape and may be as wide as 50 km (Puckridge *et al.*, 2000) to 500 km (Gibling *et al.*, 1998).

A. The Nature of Floods in Constrained Streams

In incised, constrained streams, floods are accompanied by large and rapid increases in flow volume and velocity, which generate high shear stresses on the stream bottom (Fig. 2) (Death and Winterbourn, 1995; Gordon *et al.*, 2004). The levels of suspended solids in the water column may rise prodigiously with larger particles being carried the higher the velocity of water flow. Indeed, in steep areas with loose soil, heavy rain may generate hyperconcentrated floods in which sediment concentrations may comprise 20–47% by volume (Costa, 1988).

The Flood in the Darling, 1890. W.C. Piguenit

Figure 1   The flood in the Darling by W.C. Piguenit 1890. This painting came to represent a symbol of the vast floodwaters that inundated floodplains but might be stored and made available for irrigation in the dry inland catchments. © Art Gallery of New South Wales. (See Color Plate 9.)

Floods in constrained streams may dramatically change the morphology of the streambed, banks, and riparian zone. The entrainment and transport of sediments leads to major changes in the streambed morphology, with some patches being scoured out and others being filled in (Matthaei *et al.*, 1999a), habitats may be depleted or removed and new ones created (O'Connor and Lake, 1994; Costa and O'Connor, 1995). Floods may sweep away algae, macrophytes, benthic fauna, and riparian plants along with sediments, benthic organic matter, and coarse wood (Fisher *et al.*, 1982; Molles, 1985; Brooks, 1997; Lytle, 2000). In large constrained rivers with wide channels, floods may move and reposition sediments, coarse wood, and riparian plants that generate new islands while depleting others (Gurnell *et al.*, 2005). Accumulations of branches and leaves form debris dams, which increase retention of organic matter, sediments, and nutrients, provide food in the form of detritus, and may be created by small floods and spates, but, with big floods, may be swept away.

Floods may even change stream type. For example, floods are often responsible for triggering erosion following land clearing and mining activities, generating large volumes of sediments that can inundate the streambed, completely modifying channel morphology (Gilbert, 1917; Rutherfurd, 1996; Davis and Finlayson, 2000; Prosser *et al.*, 2001). In particular situations, this process can affect long lengths of stream channel creating "sand slugs" (Rutherfurd, 1996;

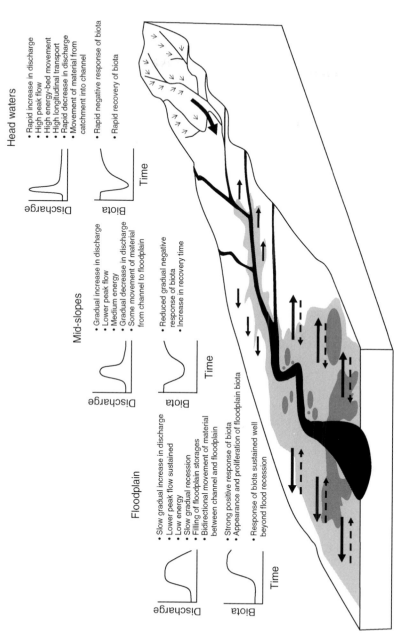

Head waters

- Rapid increase in discharge
- High peak flow
- High energy-bed movement
- High longitudinal transport
- Rapid decrease in discharge
- Movement of material from catchment into channel

- Rapid negative response of biota
- Rapid recovery of biota

Mid-slopes

- Gradual increase in discharge
- Lower peak flow
- Medium energy
- Gradual decrease in discharge
- Some movement of material from channel to floodplain

- Reduced gradual negative response of biota
- Increase in recovery time

Floodplain

- Slow gradual increase in discharge
- Lower peak flow sustained
- Low energy
- Slow gradual recession
- Filling of floodplain storages
- Bidirectional movement of material between channel and floodplain

- Strong positive response of biota
- Appearance and proliferation of floodplain biota
- Response of biota sustained well beyond flood recession

Figure 2 Diagrammatic representation of the changing nature of the flood hydrograph and the biotic response to flooding as one moves from small upland streams to large lowland rivers. Solid arrows depict the longitudinal and lateral movement of water as floodwaters move down the system. Dashed arrows indicate the movement of water, materials, and energy during flood recession.

Bartley and Rutherfurd, 2005), which become quite static, thus permanently altering the geomorphic characteristics of streams.

In stony upland streams, the effects of floods and spates on the streambed have been studied by using marked stones that are tracked before and after floods (Townsend et al., 1997b; Downes et al., 1998; Matthaei et al., 2004). With this approach, Downes et al. (1998) working in the Acheron River system, a tributary of the Goulburn River, Victoria, found that there was considerable movement and burial of stones with floods and spates, and that floods generated new patterns of patchiness on the streambed. Overall, in upland gravel-bed streams the movement and burial of stones (Downes et al., 1998), the patchy scouring and infilling (Matthaei et al., 1999a,b), the flushing away of fine sediments, and the abrasion of surfaces (Milner et al., 1981; Bond and Downes, 2003; Webb et al., 2006) by floods create a dynamic mosaic of patches subjected to varying degrees of localized disturbance (Lancaster and Hildrew, 1993). In sand-bed streams, which usually have a gentler gradient than gravel-bed streams, floods can move large amounts of sand, reshaping the channel, creating large patches of scour and fill (Fisher et al., 1982; O'Connor and Lake, 1994; Palmer et al., 1996).

B. Floods in Unconstrained Floodplain Rivers

In low-gradient lowland rivers the nature and effects of floods are markedly different from those in upland reaches. Floodplain systems usually have a meandering channel and a wide floodplain with old cutoff meanders forming billabongs or oxbow lakes. The channel bed and banks consist of fine sediments—sands, silts, and clays—and the channel in the natural state contains large amounts of coarse wood or snags (Gippel et al., 1996). With increases in flow, the low gradient means that velocities rise much less dramatically than in upland streams, and the increase in depth is much more pronounced. With major floods, water moves out onto the floodplain replenishing the billabongs, anabranches, flood runners, and temporary wetlands. The flood may move large amounts of organic matter both into and out of the channel (Fig. 2), but generally the large logs in the channel making up snags remain stable (Golladay, 1997; Nicol et al., 2001). Such floods occur as predictable seasonal events—the flood pulse (Junk et al., 1989; Tockner et al., 2000). However, extreme floods, such as a one in a 100-year flood, with very high volumes may generate shear forces both in the channel and across the floodplain that riparian vegetation and sediments are removed (Parsons et al., 2005). Such rare events create an intricate platform for subsequent succession. Flood pulses occur in many floodplain rivers such as those of the southern Murray–Darling Basin (e.g., Murray, Ovens, Goulburn, Murrumbidgee). In contrast, the dryland rivers of inland Australia (e.g., Darling, Paroo), experience large and infrequent floods that are highly unpredictable on an interannual basis (Puckridge et al., 2000), but when they occur thousands of square kilometers may be inundated (Bunn et al., 2006) and distant and isolated waterholes are connected (Arthington et al., 2005; Bunn et al., 2006). Overall, floods in lowland unconstrained rivers perform the vital function of greatly augmenting both lateral and longitudinal connectivity, which is of critical importance ecologically (Junk et al., 1989; Bunn and Arthington, 2002).

IV. ECOLOGICAL RESPONSES TO FLOODS

A. Adaptations

In streams and rivers, floods constitute a major force structuring communities and ecosystems. Floods and other disturbances, such as droughts, are strong selection forces on the biota, molding adaptations and traits (morphological, life history, functional, behavioral) so that the flora and fauna can survive and persist under the prevailing disturbance regime (Lytle and Poff, 2004). Not surprisingly, the nature of these adaptations also mirrors the longitudinal changes in the physical characteristic of floods.

In upland streams, the selection pressures are likely to be focused on dealing with high-energy, short-durational, flow events by either resistance or resilience (avoidance and recovery). In a study of invertebrates in a New Zealand upland river, Townsend et al. (1997a) found that traits associated with resistance (small body size, high adult mobility, wide habitat tolerance) and with resilience (substrate clinging, two or more life stages outside the river) were positively correlated with the intensity of flooding. In comparison, adaptations of floodplain biota center on maximizing benefits during floods and on avoidance strategies when the floodplain dries up. Aquatic invertebrates dwelling on floodplains are thus characterized by traits associated with rapid growth and reproduction during inundation such as small body size, high adult mobility, and generalized diets. Between floods, many organisms withstand prolonged dry periods with desiccation-resistant life history stages, such as the egg banks of microinvertebrates (Boulton and Lloyd, 1991; Jenkins and Boulton, 2003) or the seed banks of many floodplain plants (Capon, 2005), while others, such as winged invertebrates, are highly mobile and presumably able to seek other water sources.

Avoidance strategies—often involving refugia—are crucial ways biota both resist and recover from floods (Sedell et al., 1990; Lancaster and Belyea, 1997). Refugia can be defined as habitats or environmental factors that coupled with morphological, life history, and behavioral attributes of animals reduce the impacts of disturbance (Sedell et al., 1990; Lancaster and Belyea, 1997). Refugia can greatly reduce mortality during disturbances, and can thus strongly influence subsequent patterns of biotic recovery (resilience; Sedell *et al.*, 1990). Refugia used by animals during floods are many. They include moving behind stable stones (Matthaei *et al.*, 2004) and behind wood and debris dams (Palmer *et al.*, 1996), moving into areas of low shear stress, such as pools and the sides of streams (Lancaster and Hildrew, 1993; Brooks, 1997; Fausch *et al.*, 2002), moving down into the hyporheic zone, as suggested by Marchant (1988) in the Acheron River, and leaving the stream entirely (Lytle and Poff, 2004). A startling example of refugial behavior and habitat shift is shown by belostomatid bugs in US desert streams, whereby triggered by increased rainfall heralding a flash flood, they fly out of the stream, returning when the flood abates (Lytle, 1999). Spatial and temporal variation in the availability and accessibility of different types of refugia, and the way in which these are exploited by individual species, may also lead to very different recovery trajectories following individual floods.

B. Constrained Streams: Effects and Recovery

In general, it appears that the biota of constrained, upland streams have a low resistance but a strong resilience to floods. The high flow velocities and shear stresses, so important to the physical processes occurring during floods sweep away and damage micro- and macro-plants and animals, abrade and remove attached biota (e.g., biofilms), and damage riparian vegetation. Usually patches in the stream are greatly depleted of biota and mortality is assumed to be high, although in reality the fate of organisms lost from the streambed is poorly understood (Poff, 1992; Bond and Downes, 2003). On the positive side, floods are major generators of longitudinal connectivity allowing animals such as fish to move up and down the stream (David and Closs, 2002). Increased downstream drift of invertebrates—catastrophic drift—due to animals being dislodged may occur (Anderson and Lehmkuhl, 1968; Bond and Downes, 2003), though during floods in the Lerderderg River, Victoria, Brooks (1997) found that drift decreased. It seems that in this latter case, the invertebrates sought refugia under bottom stones. The extent of damage done by floods depends on the peak volume, flashiness, and duration (Molles, 1985; Grimm and Fisher, 1989; Maier, 2001). Unpredictable floods can be more damaging than predictable, seasonal ones (Giller *et al.*, 1991; Lytle and Poff, 2004).

Floods generate patches of areas disturbed to varying extents. In gravel-bed streams, patches may range from a stone that has been turned over to large scoured-out and filled-in areas (Downes *et al.*, 1998; Matthaei *et al.*, 2004) that may even encompass the entire streambed (Molles, 1985). In sand-bed streams there can be great changes in the streambed with sand bars and dunes being both destroyed and created (Fisher *et al.*, 1982), and both organic matter and macroalgae being scoured out and transported downstream.

The flood in the Darling, 1890. W.C. Piguenit.

Plate 9

Within a biotic group (e.g., macroinvertebrates), floods may exert quite varying effects. For example, a flood in a perennial stream did not deplete caenid nymphs dwelling on stable substrate, whereas chironomid larvae dwelling on loose sand were greatly depleted (Miller and Golladay, 1996). Further Hendricks *et al.* (1995) found that a one in 60-year flood greatly depleted populations of three caddis larvae and one ephemeropteran, whereas there was no population change in another mayfly species (*Ephoron leucon*). Floods can have quite different effects across major components. For example, a one in a 100-year flood depleted early successional riparian vegetation, but not older successional riparian vegetation or the benthic invertebrates (Hering *et al.*, 2004).

Despite the low resistance of biota to floods in constrained streams, their recovery (resilience) is rapid and is a process of secondary succession. Stream biofilm algae, notably diatoms, rapidly colonize sand, stones, and logs (Fisher *et al.*, 1982; Steinman and McIntire, 1990), and algal colonization in some cases leads to colonization by filamentous algae (Fisher *et al.*, 1982). The high mobility and refugia use of many stream animals, from invertebrates to fish, allows them to rapidly recolonize denuded habitats (Lake, 2000). In temporary streams aerial colonization may be significant (Gray and Fisher, 1981; Lytle and Poff, 2004). Filter feeders, such as blackfly (Simuliidae) larvae, are early colonists, to be followed by highly mobile scraper-grazers such as mayfly (Ephemeroptera) nymphs. These are usually followed by more sedentary animals such as case-dwelling and net-spinning caddis larvae (Trichoptera). Predatory species are often among the last to recolonize. All of this recolonization is fairly rapid—a matter of weeks or a few months—usually a time shorter than the generation time of many of the animals (Lake, 2000). Exceptions include some slow-growing plants, such as bryophytes, which may be extremely slow to recolonize after disturbance, and which may be excluded entirely from frequently disturbed habitat patches (Muotka and Virtanen, 1995; Downes *et al.*, 2003).

Not surprisingly, floods can alter interspecific interactions, such as competition by favoring one competitor as opposed to other species (Hemphill and Cooper, 1983) and predation by affecting predators more severely than their prey (Thomson, 2002). Floods may also alter the success of invasion by exotic species. In some desert streams of southwestern United States, the native sonoran topminnow (*Poeciliopsis occidentalis*) is depleted by predation from the exotic mosquito fish (*Gambusia affinis*). However, in those streams with flash flooding, the topminnow survives as the flash flooding greatly depletes the mosquito fish (Meffe, 1984).

Floods may disrupt ecological processes in constrained streams. In Sycamore Creek, Arizona, floods greatly diminished both ecosystem primary production and respiration (Fisher *et al.*, 1982) and subsequently both showed a rapid post-flood recovery to the original autotrophic state in which the production of carbon exceeded that used in respiration ($P/R \sim 1.4–1.5$). In two rocky European streams, floods greatly reduced primary production (~37% and 53%) but did not alter respiration so drastically (~14% and 24%) (Uehlinger, 2000). Prior to flooding, the streams were autotrophic ($P/R > 1$), flooding caused them to become heterotrophic and recovery back to the pre-flood state took some time (months).

Clearly, one would expect trophic web structure to be greatly altered by floods, but this facet of ecology has been neglected in upland streams. In streams there are trophic prey subsidies between the biota of the stream and surrounding riparian zone. Insects emerging from streams are prey to terrestrial predators, such as spiders and birds, and insects falling into streams are prey for stream predators (Nakano and Murakami, 2001; Baxter et al., 2005). One would expect these subsidies to be disrupted by floods as the stream biota in particular may be both depleted and altered in composition.

It is very evident that floods can exert a very strong influence on the ecology of constrained streams. Floods may reconfigure residential habitat and refugia, deplete populations of flora and fauna, change species composition and the strength and nature of interspecific interactions, and greatly alter ecosystem processes and possibly trophic structure. Many of these alterations are steadily reversed after floods, but some effects may linger on. Large floods in particular may have lasting effects on fish (Coon, 1987; Jowett and Richardson, 1989), invertebrates (Giller et al., 1991), and riparian communities (Katz et al., 2005).

C. Response to Floods in Floodplain River Systems

As floods move down unconstrained rivers, water levels rise rapidly in the channel until the banks are overtopped. In places there may be natural levees and the water may move out onto the floodplain only at particular points. As water goes out onto the floodplain, as a slow-moving flood, (the pulsing sheetflow; Dong, 2006) the various water bodies—billabongs, temporary and permanent wetlands, channels, distributaries, flood runners—become connected, generating a great diversity of water bodies—a shifting mosaic (Ward, 1989; Amoros and Bornette, 2002; Tockner et al., 2002). The floods may be regular seasonal events (Tockner et al., 2000) as occurred in the Murray River in spring prior to river regulation (Maheshwari et al., 1995), or unpredictable, irregular events as occurs in the dryland rivers of inland Australia (Puckridge et al., 1998, 2000). Small floods or flow pulses (Puckridge et al., 1998) occur within the channel of floodplain rivers. These smaller floods can perform an important geomorphic role by maintaining features such as bars and benches which, by creating backwaters, increase hydraulic diversity and provide important habitats and disturbance refugia (Matthews, 1999; King, 2004).

As overbank floods slowly move across the floodplain, nutrients, sediment, organic matter, and biota are carried across the floodplain (Ward, 1989) stimulating a variety of primary and secondary production processes both on the floodplain itself and within off-channel habitats such as billabongs (Robertson and Bunn, 1999). Newly hatched microinvertebrates may be carried out to flooded wetlands (Jenkins and Boulton, 2003), and as flood-plain soils are inundated, nutrients and dissolved organic matter are released into the water column stimulating bacterial, phytoplankton, and microfaunal production (O'Connell et al., 2000; Valett et al., 2005). In turn, macrophyte biomass and production increase, attaining very high levels in many flood-plains (Hamilton et al., 1997; Valett et al., 2005). This boost in production is accompanied by dramatic increases in aquatic biodiversity, and in their expansion phase (sensu Stanley et al., 1997) floods temporarily transform natural floodplains into one of the most productive environments on earth (Tockner et al., in press). Thus, floods on floodplains are critical ecosystem expansion–contraction events (Stanley et al., 1997) connecting the floodplain to the river channel. This is very evident in dryland rivers when inundated floodplains become hotspots of production, with booms of aquatic faunal production attracting large populations of waterbirds (Kingsford and Norman, 2002; see also chapter by Aldo Poiani). Across inland Australia, the floods are rather unpredictable both in time and place. However, waterbirds can deal with this unpredictability as they migrate great distances to move from flood to flood (Kingsford et al., 1999; Roshier et al., 2002).

Accompanying the stimulated production, floodplain respiration also increases dramatically. Particulate detritus, such as leaves, dissolved organic matter, and nutrients from sediments, may feed production in forested flood-plains of seasonally flooding rivers (Junk et al., 1989; Ward, 1989; Tockner et al., 2000). However, across the floodplain, stagnant water bodies may occur and they may become deoxygenated (Hamilton et al., 1997) and subsequently bereft of aquatic fauna. Leachates, containing polyphenols, from inundated leaves, such as those of river red gum (Eucalyptus camaldulensis) (O'Connell et al., 2000), may build up to produce deoxygenation and "blackwater events," which can kill both invertebrates and fish (Gehrke, 1991; Gehrke et al., 1993). Such events have become more common with the advent of controlled floods (environmental flows) below dams in which attempts are made to minimize the volume of water required to inundate selected parts of the floodplain, thereby creating stagnant floods that produce blackwater events such as in Barmah Forest in 2001 (Howitt et al., 2004; King, 2005).

Water may leave the floodplain by evapotranspiration or by receding back into the channel. With the recession, nutrients, dissolved organic matter, bac-teria, micro- and macroinvertebrates and fish may move back into the river channel (Williams et al., 1998), although in comparison with the ecology of the flood expansion onto the floodplain, our ecological understanding of the recession is rather rudimentary. On the land, the boom in aquatic production is followed by a pulse of terrestrial production as the floodwaters recede. Floods by recharging soil water are vital for many riparian plants, as exemplified by river red gum (Bren and Gibbs, 1986) and other riparian species (Capon, 2005). For much of the terrestrial biota, floods act as ecological disturbances causing the death of many organisms, both plants and animals, and forcing more mobile taxa to seek refugia—either in trees or on higher parts of the floodplain (Ballinger et al., 2005, see also chapter by Andrea Ballinger and Ralph Mac Nally). However some terrestrial arthropods, such as carabid beetles and spiders, are favored by floods as they are predators of aquatic invertebrates in the receding floodwaters (Ballinger and Mac Nally, 2005). Floods by stimulating high production of aquatic invertebrates, such as flying adult insects, may be a very rich subsidy of limited duration to terrestrial predators ranging from spiders to birds and bats (Ballinger and Mac Nally, 2005; Ballinger et al., 2005). Similarly, inundation allows fish to exploit newly created floodplain habitat and take advantage of a wider array of prey items (Arthington et al., 2005). In some cases, flooding may mediate the dependence of lowland fish species on terrestrial dietary sources (Arthington et al., 2005). The importance of subsidies between terrestrial and aquatic within floodplain ecosystems warrants further attention (Ballinger and Lake, 2006).

The importance and consequences of floodplain inundation has been a major focus of flood research in lowland rivers; however, the responses of biota within the main channel are less well understood. Floods are often regarded as an important spawning trigger for lowland fish species with adults moving onto the floodplain and into channel habitats to spawn (Wellcome, 1985; Bayley, 1995). In the Murray–Darling Basin, floods were widely believed to trigger spawning in a large number of species, but much of the evidence for this was anecdotal or recorded from hatcheries rather than natural channels. More recent reviews (Humphries et al., 1999; Graham and Harris, 2005) and empirical studies (King, 2002) suggest that among native species, only a small number (e.g., golden perch, *Macquaria ambigua* and silver perch, *Bidyanus bidyanus*) rely on flooding for successful spawning, with many others spawning irrespective of flooding, often in late summer when floods rarely occur. It has also been suggested that, even in the absence of a direct spawning response to flooding, the stimulation of secondary production by floods may still be critical in increasing prey availability for the larvae of some Australian lowland fish species (Gehrke and Harris, 1994; Humphries et al., 1999; Balcombe et al., 2005).

V. THE HUMAN DIMENSION

The frequency, magnitude, and timing of floods in many rivers have been altered by human activities associated with flood mitigation, the extraction of water for irrigation, stock and domestic use, and the generation of hydroelectric power. In the Murray–Darling Basin, irrigation development first began in the late 1800s and was particularly intensive during the 1920s–1940s with the construction of Hume Dam on the Murray River at Albury (Longitude 147.0328°E, Latitude 36.1092°S), Burrinjuck on the Murrumbidgee River near Yass (Longitude 148.5875°E, Latitude 35.0039°S) as well as numerous weirs and diversion structures that assist in controlling the diversion of water into irrigation networks. Today there are more than 80 dams over 10,000-ML capacity within the Murray–Darling Basin (ANCOLD, 1990).

In the upland sections of the basin, many of the smaller streams and rivers have been dammed for the purposes of generating hydroelectric power. In most cases, these dams are relatively small and the frequency of large natural floods is largely unaffected. Such dams can however increase the overall frequency of minor floods (spates) via hydropeaking, in which flows are rapidly increased during power generation. Together with the other impacts associated with dam development, the occurrence of rapid rises in the hydrograph from hydropeaking can deplete fauna immediately below hydroelectric dams (Cushman, 1985), much as described for natural floods in constrained upland channels. The ecological impacts of both natural and anthropogenic floods in constrained systems have also been exacerbated by the loss of refuge habitats (e.g., those associated with fallen timber and backwaters) destroyed by channelization and snag removal (Negishi and Richardson, 2003; Bond and Lake, 2004)—misguided attempts at improving the flood conveyance capacity of rivers (Gippel, 1995; Gippel *et al.*, 1996; Erskine and Webb, 2003).

While flooding patterns have been altered in the headwaters and constrained reaches of many rivers, it is the lowland floodplain sections in which changes to flow regimes and patterns of flooding are most pronounced and also most widespread. In the mid to lower sections of the Murray River, for example, the effects of river regulation and water extraction have been the cause of a significant overall reduction in discharge—estimated at around 56% of the long-term mean annual discharge in the lower sections (Maheshwari *et al.*, 1995). The large storage capacity of the Hume Dam also means that flooding flows are now much less frequent, with only large floods (with a recurrence interval >20 years) capable of overwhelming the storage capacity of the system moving downstream unabated. Annual spring floods historically generated by snow-melt are now much reduced in magnitude.

While these smaller floods typically would inundate smaller areas of floodplain, they play a critical role in replenishing off-channel habitats such as billabongs and wetlands, and also in maintaining the health of large tracts of river red gum floodplain forest (Bren and Gibbs, 1986). Reductions in flooding frequency can lead to dramatic changes in the extent and health of floodplain forest, in some cases promoting tree expansion into wetland habitat (Bren, 1992), while in others causing widespread tree mortality due to water stress (Chesterfield, 1986; MDBC, 2003). Restoring floodplain forests however may not be a simple case of reinstating a more natural flooding regime. Meeson *et al.* (2002) have shown that reductions in flood frequency may exacerbate the damaging influence of livestock on red gum recruitment. In their study, heavily grazed floodplains that received less flooding had higher rates of ant seed predation, thus further reducing potential recruitment.

In floodplain sections of the river, the reduction in flood magnitude and frequency is further exacerbated by the construction of levees and other structures to control the movement of water onto and off the floodplain. In many cases, levees isolate off-channel habitats even where floods of sufficient size to inundate them still occur (Mussared, 1997). The considerable cost of maintaining levees together with the fact that, once breached, they can significantly increase the economic damage done by floods (Hickey and Salas, 1995), means there is an emerging trend toward nonstructural flood mitigation measures (Changnong, 1993; Hickey and Salas, 1995).

The elimination of floods by river regulation is a press disturbance that can have long-lasting deleterious effects. For example, in southwestern US streams, floods are critical for recruitment of important riparian tree such as native cottonwoods. However, river regulation by eliminating such floods hampers cottonwood recruitment and allows invasion by exotic tamarisk, that are well adapted to the prevailing altered regime of reduced stream flow and drier riparian soils (Rood *et al.*, 2005). The ecological effects of flow regulation are only now beginning to be fully understood. In many rivers, efforts are underway to reinstate "environmental flows"—flows that mimic components of the natural flow regime, including floods, in the hope of restoring important ecological processes. In the Snowy River, for example, flows are being returned to 28% of the natural mean annual flow (formerly 1%) (Miller, 2005). In the Murray River, efforts are also underway to reinstate the flows necessary to inundate floodplains in an attempt to restore the health of river red gum forests in selected areas (MDBC, 2003). While the preceding discussion indicates the ecological importance of floods, considerable research is required before synthetic flow regimes—that is those designed to mimic particular flow regime components—will be successful in stimulating many of the physical and biological processes associated with natural floods (Bunn and Arthington, 2002; Graham and Harris, 2005).

VI. CONCLUSIONS

High water events, floods, as disturbances are a major force shaping the ecology of streams and rivers. In constrained streams, usually in upland areas near the source, floods are marked by their high power—high velocities and shear stress. Such floods can change the shape of the channel, moving and scouring sediments, degrading and creating habitat, and removing and damaging biota.

The biota of such streams generally have a low resistance to the floods and a high resilience—the capacity to recover. In most cases but certainly not all, the recovery by biota is relatively rapid. Such recovery occurs through recolonization from refugia and unaffected tributaries followed by subsequent succession. Floods in lowland rivers with very low slopes inundate floodplains. Such floods may be very extensive and of long duration. They may be also a destructive disturbance to terrestrial biota, but are vital replenishers to the aquatic and riparian biota. With the advent of water, nutrients are released, desiccation-resistant plants and fauna emerge and grow, resulting in a pulse, a boom of primary and secondary production. Migrating fish and birds may breed. As the floodwaters recede, they carry nutrients, detritus, and biota back into the river channel, stimulating further production.

Human activities have greatly degraded rivers, with the loss of floods being a major contributor to this degradation. Loss of connectivity by barriers, such as through dams, often coupled with water extraction, has reduced flood frequency and magnitude. In constrained rivers, lack of floods may lead to lack of sediment, colmation of the streambed and loss of downstream transport of nutrients, detritus, and biota. Loss of flood inundation of floodplains has occurred through the effects of upstream dams, but has also been produced by provision of flood protection measures, such as the building of levees, floodplain harvesting, and water diversions for irrigated agriculture.

ACKNOWLEDGMENTS

This work was supported by Land and Water Australia (Sam Lake), the Murray–Darling Basin Commission (Paul Reich), and eWater CRC (Nick Bond). We would also like to thank Elise King for assistance in preparing the chapter, Aldo Poiani for constructive comments on an earlier draft of the chapter, and the Art Gallery of New South Wales for granting permission to publish Fig. 1.

REFERENCES

Amoros, C. and Bornette, G. (2002) Connectivity and biocomplexity in waterbodies of riverine floodplains. *Freshw. Biol.* **47**, 761–776.
ANCOLD (1990) "Register of Large Dams in Australia." Australian National Committee on Large Dams, c/o Hydro-Electric Commission, Hobart.
Anderson, N.H. and Lehmkuhl, D.M. (1968) Catastrophic drift of insects in a woodland stream. *Ecology* **49**, 198–206.
Arthington, A.H., Balcombe, S.R., Wilson, G.A., Thoms, M.C. and Marshall, J. (2005) Spatial and temporal variation in fish-assemblage structure in isolated waterholes during the 2001 dry season of an arid-zone floodplain river, Cooper Creek, Australia. *Mar. Freshw. Res.* **56**, 25–35.

Balcombe, S.R., Bunn, S.E., McKenzie-Smith, F.J. and Davies, P.M. (2005) Variability of fish diets between dry and flood periods in an arid zone floodplain river. *J. Fish Biol.* **67**, 1552–1567.

Ballinger, A. and Lake, P.S. (2006) Energy and nutrient fluxes from rivers and streams into terrestrial food webs. *Mar. Freshw. Res.* **57**, 15–28.

Ballinger, A. and Mac Nally, R. (2005) Flooding in Barmah-Millewa forest: Catastrophe or opportunity for non-aquatic fauna? *Proc. R. Soc. Victoria* **117**, 111–115.

Ballinger, A., Mac Nally, R. and Lake, P.S. (2005) Immediate and longer-term effects of managed flooding on floodplain invertebrate assemblages in south-eastern Australia: Generation and maintenance of a mosaic landscape. *Freshw. Biol.* **50**, 1190–1205.

Bartley, R. and Rutherfurd, I. (2005) Measuring the reach-scale geomorphic diversity of streams: Application to a stream disturbed by a sediment slug. *River Res. Appl.* **21**, 39–59.

Baxter, C.V., Fausch, K.D. and Saunders, W.C. (2005) Tangled webs: Reciprocal flows of invertebrate prey link streams and riparian zones. *Freshw. Biol.* **50**, 201–220.

Bayley, P.B. (1995) Understanding large river floodplain ecosystems. *Bioscience* **45**, 153–158.

Bond, N.R. and Downes, B.J. (2003) The independent and interactive effect of fine sediment and flow on benthic invertebrate communities characteristic of upland streams. *Freshw. Biol.* **48**, 455–465.

Bond, N.R. and Lake, P.S. (2004) Disturbance regimes and stream restoration: The importance of restoring refugia. In: *Proceedings of the 4th Australian Stream Management Conference* (Ed. by I. Rutherfurd, I. Wiszniewski, M. Askey-Doran and R. Glazik), pp. 90–94. Department of Primary Industries, Water and Environment, Hobart, Tasmania.

Boulton, A.J. and Lloyd, L.N. (1991) Macroinvertebrate assemblages in floodplain habitats of the lower river Murray, South Australia. *Regul. Rivers* **6**, 183–201.

Bren, L.J. (1992) Tree invasion of an intermittent wetland in relation to changes in the flooding frequency of the river Murray, Australia. *Aust. J. Ecol.* **17**, 395–408.

Bren, L.J. and Gibbs, N.L. (1986) Relationships between flood frequency, vegetation and topography in a river red gum forest. *Aust. For. Res.* **16**, 357–370.

Brooks, S. (1997) Impacts of flood disturbance on the macroinvertebrate assemblages of an upland stream. Ph.D. Thesis, School of Biological Sciences, Monash University, Melbourne, Australia.

Bunn, S.E. and Arthington, A. (2002) Basic principles and ecological consequences of altered flow regimes for aquatic biodiversity. *Environ. Manag.* **30**, 492–507.

Bunn, S.E., Thoms, M.C., Hamilton, S.K. and Capon, S.J. (2006) Flow variability in dryland rivers: Boom, bust and the bits in between. *River Res. Appl.* **22**, 179–186.

Capon, S.J. (2005) Flood variability and spatial variation in plant community composition and structure on a large and floodplain. *J. Arid Environ.* **60**, 283–302.

Changnong, S. (1993) The 1993 flood's aftermath: Risks, root causes, and lessons for the future. *J. Contemp. Water Res. Ed.* **130**, 70–74.

Chesterfield, E.A. (1986) Changes in the vegetation of the river red gum forest at Barmah, Victoria. *Aust. For.* **49**, 4–15.

Coon, T.G. (1987) Response of benthic riffle fishes to variation in stream discharge and temperature. In: *Community and Evolutionary Ecology of North American Stream Fishes* (Ed. by W.J. Matthews and D.S. Heins), pp. 77–85. University of Oklahoma, Oklahoma.

Costa, J.E. (1988) Rheologic, geomorphic, and sedimentologic differentiation of water floods, hyperconcentrated flows, and debris flows. In: *Flood Geomorphology* (Ed. by V.R. Baker, R.C. Kochel and P.C. Patton), pp. 113–122. John Wiley & Sons, New York.

Costa, J.E. and O'Connor, J.E. (1995) Geomorphically effective floods. In: *Natural and Anthropogenic Influences in Fluvial Geomorphology* (Ed. by J.E. Costa, J.E. Miller, A.J. Potter and P.R. Wilcock), pp. 45–56. American Geophysical Union, Washington, DC.

Cushman, R.M. (1985) Review of ecological effects of rapidly varying flows downstream from hydroelectric facilities. *N. Am. J. Fish. Manag.* 5, 330–339.

David, B.O. and Closs, G.P. (2002) Behavior of a stream-dwelling fish before, during, and after high-discharge events. *Trans. Am. Fish. Soc.* 131, 762–771.

Davis, J.A. and Finlayson, B. (2000) *Sand Slugs and Stream Degradation: The Case of the Granite Creeks, North-East Victoria.* Cooperative Research Centre for Freshwater Ecology, Canberra.

Death, R.G. and Winterbourn, M.J. (1995) Diversity patterns in stream benthic invertebrate communities: The influence of habitat stability. *Ecology* 76, 1446–1460.

Dong, Q. (2006) Pulsing sheetflow and wetland integrity. *Front Ecol. Environ.* 4, 9.

Downes, B.J., Lake, P.S., Glaister, A. and Webb, A. (1998) Scales and frequencies of disturbances: Rock size, bed packing and variation among upland streams. *Freshw. Biol.* 40, 625–639.

Downes, B.J., Entwisle, T. and Reich, P. (2003) Effects of flow regulation on disturbance frequencies and in-channel bryophytes and macroalgae in some upland streams. *River Res. Appl.* 19, 27–42.

Erskine, W.D. and Webb, A. (2003) Desnagging to resnagging: New directions in river rehabilitation in southeastern Australia. *River Res. Appl.* 19, 233–249.

Fausch, K.D., Torgersen, C.E., Baxter, C.V. and Li, H.W. (2002) Landscapes to riverscapes: Bridging the gap between research and conservation of stream fishes. *Bioscience* 52, 483–498.

Fisher, S.G., Gray, L.J., Grimm, N.B. and Busch, D.E. (1982) Temporal succession in a desert stream ecosystem following flash flooding. *Ecol. Monogr.* 52, 93–110.

Gehrke, P.C. (1991) Avoidance of inundated floodplain habitat by larvae of golden perch (*Macquaria ambigua* Richardson): Influence of water quality or food distribution? *Aust. J. Mar. Freshw. Res.* 42, 265–279.

Gehrke, P.C. and Harris, J.H. (1994) The role of fish in cyanobacterial blooms in Australia. *Aust. J. Mar. Freshw. Res.* 45, 905–915.

Gehrke, P.C., Revell, M.B. and Philbey, A.W. (1993) Effects of river red gum, *Eucalyptus camaldulensis*, litter on golden perch, *Macquaria ambigua*. *J. Fish Biol.* 43, 265–279.

Gibling, M.R., Nanson, G.C. and Maroulis, J.C. (1998) Anastomosing river sedimentation in the channel country of central Australia. *Sedimentology* 45, 595–619.

Gilbert, G.K. (1917) *Hydraulic-Mining Debris in the Sierra Nevada.* Department of the Interior, United States Geological Survey, Washington.

Giller, P.S., Sangpradub, N. and Twomey, H. (1991) Catastrophic flooding and macroinvertebrate structure. *Internat. Verein. Theor. Angew. Limnol.* 24, 1724–1729.

Gippel, C.J. (1995) Environmental hydraulics of large woody debris in streams and rivers. *J. Environ. Eng.* 121, 388–395.

Gippel, C.J., Finlayson, B.L. and O'Neill, I.C. (1996) Distribution and hydraulic significance of large woody debris in a lowland Australian river. *Hydrobiologia* **318**, 179–194.

Golladay, S.W. (1997) Suspended particulate organic matter concentration and export in streams. *J. N. Am. Benthol. Soc.* **16**, 122–131.

Gordon, N.D., McMahon, T.A., Finlayson, B.F., Gippel, C.J. and Nathan, R.J. (2004) *Stream Hydrology: An Introduction for Ecologists.* Wiley, Chichester.

Graham, R. and Harris, J.H. (2005) *Floodplain Inundation and Fish Dynamics in the Murray-Darling Basin. Current Concepts and Future Research: A Scoping Study.* Cooperative Research Centre for Freshwater Ecology, Canberra.

Gray, L.J. and Fisher, S.G. (1981) Postflood recolonisation pathways of macro-invertebrates in a lowland Sonoran desert stream. *Am. Midl. Nat.* **106**, 249–257.

Grimm, N.B. and Fisher, S.G. (1989) Stability of periphyton and macroinvertebrates to disturbance by flash floods in a desert stream. *J. N. Am. Benthol. Soc.* **8**, 293–307.

Gurnell, A., Tockner, K., Edwards, P. and Petts, G. (2005) Effects of deposited wood on biocomplexity of river corridors. *Front. Ecol. Environ.* **3**, 377–382.

Hamilton, S.K., Sippel, S.J., Calheiros, D.F. and Melack, J.M. (1997) An anoxic event and other biogeochemical effects of the Pantanal wetland on the Paraguay River. *Limnol. Oceanogr.* **42**, 257–272.

Hemphill, N. and Cooper, S.D. (1983) The effect of physical disturbance on the relative abundances of two filter-feeding insects in a small stream. *Oecologia* **58**, 378–382.

Hendricks, A.C., Willis, L.D. and Snyder, C. (1995) Impact of flooding on the densities of selected aquatic insects. *Hydrobiologia* **299**, 241–247.

Hering, D., Gerhard, M., Manderbach, R. and Reich, M. (2004) Impact of a 100-year flood on vegetation, benthic invertebrates, riparian fauna and large woody debris standing stock in an alpine floodplain. *River Res. Appl.* **20**, 445–457.

Hickey, J.T. and Salas, J.D. (1995) Environmental effects of extreme floods. In: *United States-Italy Research Workshop on Hydrometeorology, Impacts, and Management of Extreme Floods* (Ed. by J.D. Salas and F. Siccardi). [Electronic Resource] available at http://www.engr.colostate.edu/~jsalas/us-italy/papers/33hickey.pdf

Howitt, J., Baldwin, F., Rees, G.N. and Williams, J.E. (2004) Modelling blackwater: Predicting water quality during flooding of lowland river forests. In: *Blackwater Model—A Computer Model to Predict Dissolved Oxygen and Dissolved Carbon Downstream of Barmah-Millewa Forest Following a Flood* (Ed. by J. Howitt, D.S. Baldwin and G.N. Rees), pp. 22–46. Murray-Darling Freshwater Research Centre, Albury, Australia.

Humphries, P., King, A.J. and Koehn, J.D. (1999) Fish, flows and floodplains: Links between freshwater fishes and their environment in the Murray-Darling River system, Australia. *Environ. Biol. Fishes* **56**, 129–151.

Jenkins, K.M. and Boulton, A.J. (2003) Connectivity in a dryland river: Short-term aquatic microinvertebrate recruitment following floodplain inundation. *Ecology* **84**, 2708–2727.

Jowett, I.G. and Richardson, J. (1989) Effects of a severe flood on instream habitat and trout populations in seven New Zealand rivers. *N. Z. J. Mar. Freshw. Res.* **23**, 11–18.

Junk, W.J., Bayley, P.B. and Sparks, R.E. (1989) The flood-pulse concept in river-floodplain systems. *Can. J. Fish. Aquat. Sci.* **106**, 110–127.

Katz, G.L., Friedman, J.M. and Beatty, S.W. (2005) Delayed effects of flood control on a flood-dependent riparian forest. *Ecol. Appl.* **15**, 1019–1035.

King, A. (2002) Recruitment ecology of fish in floodplain rivers of the southern Murray-Darling Basin, Australia. Ph.D. Thesis, School of Biological Sciences, Monash University, Melbourne, Australia.

King, A.J. (2004) Ontogenetic patterns of habitat use by fishes within the main channel of an Australian floodplain river. *J. Fish Biol.* **65**, 1582–1603.

King, A.J. (2005) Fish and the Barmah-Millewa forest—history, status and management challenges. *Proc. R. Soc. Victoria* **117**, 117–125.

Kingsford, R.T. and Norman, F.I. (2002) Australian waterbirds—products of the continent's ecology. *Emu* **102**, 47–69.

Kingsford, R.T., Curtin, A.L. and Porter, J. (1999) Water flows on Cooper Creek in arid Australia determine 'boom' and 'bust' periods for waterbirds. *Biol. Conserv.* **88**, 231–248.

Lake, P.S. (1990) Disturbing hard and soft bottom communities: A comparison of marine and freshwater environments. *Aust. J. Ecol.* **15**, 477–488.

Lake, P.S. (2000) Disturbance, patchiness and diversity in streams. *J. N. Am. Benthol. Soc.* **19**, 573–592.

Lancaster, J. and Belyea, L.R. (1997) Nested hierarchies and scale-dependence of mechanisms of flow refugium use. *J. N. Am. Benthol. Soc.* **16**, 221–238.

Lancaster, J. and Hildrew, A.G. (1993) Flow refugia and the microdistribution of lotic macroinvertebrates. *J. N. Am. Benthol. Soc.* **12**, 385–393.

Lytle, D.A. (1999) Use of rainfall cues by *Abedus herberti* (Hemiptera: Belostomatidae): A mechanism for avoiding flash floods. *J. Insect Behav.* **12**, 1–12.

Lytle, D.A. (2000) Biotic and abiotic effects of flash flooding in a montane desert stream. *Arch. Hydrobiol.* **150**, 85–100.

Lytle, D.A. and Poff, N.L. (2004) Adaptation to natural flow regimes. *Trends Ecol. Evol.* **19**, 94–100.

Maheshwari, B.L., Walker, K.F. and McMahon, T.A. (1995) Effects of regulation on the flow regime of the river Murray, Australia. *Regul. Rivers* **10**, 15–38.

Maier, K.J. (2001) The influence of floods on benthic insect populations in a Swiss mountain stream and their strategies of damage prevention. *Arch. Hydrobiol.* **150**, 227–247.

Marchant, R. (1988) Vertical distribution of benthic invertebrates in the bed of the Thomson River, Victoria. *Aust. J. Mar. Freshw. Res.* **39**, 775–784.

Matthaei, C.D., Peacock, K.A. and Townsend, C.R. (1999a) Scour and fill patterns in a New Zealand stream and potential implications for invertebrate refugia. *Freshw. Biol.* **42**, 41–57.

Matthaei, C.D., Peacock, K.A. and Townsend, C.R. (1999b) Patchy surface stone movement during disturbance in a New Zealand stream and its potential significance for the fauna. *Limnol. Oceanogr.* **44**, 1091–1102.

Matthaei, C.D., Townsend, C.R., Arbuckle, C.J., Peacock, K.A., Guggelberger, C., Kuster, C.E. and Huber, H. (2004) Disturbance, assembly rules, and benthic communities in running waters: A review and some implications for restoration projects. In: *Assembly Rules and Restoration Ecology. Bridging the Gap between Theory and Practice* (Ed. by V.M. Temperton, R.J. Hobbs, T. Nuttle and S. Halle), pp. 367–388. Island Press, Washington, USA.

Matthews, W.J. (1999) *Patterns in Freshwater Fish Ecology.* Chapman & Hall, New York.

McMahon, T.A., Finlayson, B.L., Haines, A.T. and Srikanthan, R. (1992) *Global Runoff: Continental Comparisons of Annual Flows and Peak Discharges.* Catena Verlag, Cremlingen-Destedt.

MDBC (2003) *Preliminary Investigations into Observed River Red Gum Decline Along the River Murray Below Euston.* Murray-Darling Basin Commission, Canberra.

Meeson, N., Robertson, A.I. and Jansen, A. (2002) The effects of flooding and livestock on post-dispersal seed predation in river red gum habitats. *J. Appl. Ecol.* **39**, 247–258.

Meffe, G.K. (1984) Effects of abiotic disturbance on coexistence of predator-prey fish species. *Ecology* **65**, 1525–1534.

Miller, A.M. and Golladay, S.W. (1996) Effects of spates and drying on macroinvertebrate assemblages of an intermittent and a perennial prairie stream. *J. N. Am. Benthol. Soc.* **15**, 670–689.

Miller, C. (2005) *Snowy River Story: The Grassroots Campaign to Save a National Icon.* ABC Books, Melbourne.

Milner, N.J., Scullion, J., Carling, P.A. and Crisp, D.T. (1981) The effects of discharge on sediment dynamics and consequent effects on invertebrates and salmonids in upland rivers. *Adv. Appl. Biol.* **6**, 153–220.

Molles, M.C. (1985) Recovery of a stream invertebrate community from a flash flood in Tesuque Creek, New Mexico. *Southwest Nat.* **30**, 279–287.

Muotka, T. and Virtanen, R. (1995) The stream as a habitat templet for bryophytes: Species' distributions along gradients in disturbance and substratum heterogeneity. *Freshw. Biol.* **33**, 141–160.

Mussared, D. (1997) *Living on Floodplains.* Cooperative Research Centre for Freshwater Ecology and the Murray-Darling Basin Commission, Canberra, Australia.

Nakano, S. and Murakami, M. (2001) Reciprocal subsidies: Dynamic interdependence between terrestrial and aquatic food webs. *Proc. Natl. Acad. Sci. USA* **98**, 166–170.

Negishi, J.N. and Richardson, J.S. (2003) Responses of organic matter and macroinvertebrates to placements of boulder clusters in a small stream of southwestern British Columbia, Canada. *Can. J. Fish. Aquat. Sci.* **60**, 247–258.

Nicol, S.J., Bearlin, A.R., Robley, A.J., Koehn, J.D. and Lieschke, J.A. (2001) Distribution of large woody debris in the mid-reaches of the Murray River. *Ecol. Manag. Restor.* **2**, 64–67.

O'Connell, M., Baldwin, D.S., Robertson, A.I. and Rees, G. (2000) Release and bioavailability of dissolved organic matter from floodplain litter: Influence of origin and oxygen levels. *Freshw. Biol.* **45**, 333–342.

O'Connor, N.A. and Lake, P.S. (1994) Long-term and seasonal large-scale disturbances of a small lowland stream. *Aust. J. Mar. Freshw. Res.* **45**, 243–255.

Palmer, M.A., Arensburger, P., Martin, A.P. and Denman, D.W. (1996) Disturbance and patch specific responses: The interactive effects of woody debris and floods on lotic invertebrates. *Oecologia* **105**, 247–257.

Parsons, M., McLoughlin, C.A., Kotschy, K.A., Rogers, K.H. and Rountree, M.W. (2005) The effects of extreme floods on the biophysical heterogeneity of river landscapes. *Front. Ecol. Environ.* **3**, 487–494.

Pickett, S.T.A. and White, P.S. (1985) *The Ecology of Natural Disturbance and Patch Dynamics.* Academic Press, New York.

Poff, N.L. (1992) Why disturbance can be predictable: A perspective on the definition of disturbance in streams. *J. N. Am. Benthol. Soc.* **11**, 86–92.

Prosser, I.P., Rutherfurd, I.D., Olley, J.M., Young, W.J., Wallbrink, P.J. and Moran, C.J. (2001) Large scale patterns of erosion and sediment transport in river networks, with examples from Australia. *Mar. Freshw. Res.* **52**, 81–99.

Puckridge, J.T., Sheldon, F., Walker, K.F. and Boulton, A.J. (1998) Flow variability and the ecology of large rivers. *Mar. Freshw. Res.* **49**, 55–72.

Puckridge, J.T., Walker, K.F. and Costelloe, J.F. (2000) Hydrological persistence and the ecology of dryland rivers. *Regul. Rivers* **16**, 385–402.

Robertson, A.I. and Bunn, S.E. (1999) Sources, sinks and transformations of organic carbon in Australian floodplain rivers. *Mar. Freshw. Res.* **50**, 813–829.

Rood, S.B., Samuelson, G.M., Braatne, J.H., Gourley, C.R., Hughes, F.M.R. and Mahoney, J.M. (2005) Managing river flows to restore floodplain forests. *Front. Ecol. Environ.* **3**, 193–201.

Roshier, D.A., Robertson, A.I. and Kingsford, R.T. (2002) Responses of waterbirds to flooding in an arid region of Australia and implications for conservation. *Biol. Conserv.* **106**, 399–411.

Rutherfurd, I. (1996) Sand slugs in SE Australian streams: Origins, distribution and management. In: *First National Conference on Stream Management in Australia* (Ed. by I. Rutherfurd and M. Walker), pp. 29–34. Cooperative Research Centre for Catchment Hydrology, Melbourne.

Rykiel, E.J. (1985) Towards a definition of ecological disturbance. *Aust. J. Ecol.* **10**, 361–365.

Sedell, J.R., Reeves, G.H., Hauer, F.R., Stanford, J.A. and Hawkins, C.P. (1990) Role of refugia in recovery from disturbances: Modern fragmented and disconnected river systems. *Environ. Manag.* **14**, 711–724.

Smith, K. and Ward, R. (1998) *Floods: Physical Processes and Human Impacts.* John Wiley & Sons, Chichester, UK.

Stanley, E.H., Fisher, S.G. and Grimm, N.B. (1997) Ecosystem expansion and contraction in streams. *Bioscience* **47**, 427–435.

Steinman, A.D. and McIntire, C.D. (1990) Recovery of lotic periphyton communities after disturbance. *Environ. Manag.* **14**, 589–604.

Thomson, J.R. (2002) The effects of hydrological disturbance on the densities of macroinvertebrate predators and their prey in a coastal stream. *Freshw. Biol.* **47**, 1333–1351.

Tockner, K., Malard, F. and Ward, J.V. (2000) An extension of the flood pulse concept. *Hydrol. Proc.* **14**, 2861–2883.

Tockner, K., Ward, J.V., Edwards, P.J. and Kollmann, J. (2002) Riverine landscapes: An introduction. *Freshw. Biol.* **47**, 497–500.

Tockner, K., Bunn, S.E., Gordon, C., Naiman, R.J., Quinn, G.P. and Stanford, J.A. (in press) Floodplains: Critically threatened ecosystems. In: *State of the World's Waters* (Ed. by N. Polunin), Cambridge University Press, Cambridge, UK.

Townsend, C.R., Doledec, S. and Scarsbrook, M.R. (1997a) Species traits in relation to temporal and spatial heterogeneity in streams: A test of habitat templet theory. *Freshw. Biol.* **37**, 367–387.

Townsend, C.R., Scarsbrook, M.R. and Dolèdec, S. (1997b) Quantifying disturbance in streams: Alternative measures of disturbances in relation to macroinvertebrate species traits and species richness. *J. N. Am. Benthol. Soc.* **16**, 531–544.

Uehlinger, U. (2000) Resistance and resilience of ecosystem metabolism in a flood-prone river system. *Freshw. Biol.* **45**, 319–332.

Valett, H.M., Baker, M.A., Morrice, J.A., Crawford, C.S., Molles, M.C., Dahm, C.N., Moyer, D.L., Thibault, J.R. and Ellis, L.M. (2005) Biogeochemical and metabolic responses to the flood pulse in a semiarid floodplain. *Ecology* **86**, 220–234.

Vannote, R.L., Minshall, G.W., Cummins, K.W., Sedell, J.R. and Cushing, C.E. (1980) The river continuum concept. *Can. J. Fish. Aquat. Sci.* **37**, 130–137.

Ward, J.V. (1989) The four-dimensional nature of lotic ecosystems. *J. N. Am. Benthol. Soc.* **8**, 2–8.

Ward, R. (1978) *Floods: A Geographical Perspective.* MacMillan, London, UK.

Webb, J.A., Downes, B.J., Lake, P.S. and Glaister, A. (2006) Quantifying abrasion of stable substrata in streams: A new disturbance index for epilithic biota. *Hydrobiologia* **559**, 443–453.

Webster, J.R., Gurtz, M.E., Hains, J.J., Meyer, J.L., Swank, W.T., Waide, J.B. and Wallace, J.R. (1983) Stability of stream ecosystems. In: *Stream Ecology. Application and Testing of General Ecological Theory* (Ed. by J.R. Barnes and G.W. Minshall), pp. 355–395. Plenum Press, New York.

Wellcome, R.L. (1985) *River Fisheries*. FAO Technical Bulletin 262. FAO, Rome.

Williams, W.D., De Deckker, P. and Shiel, R.J. (1998) The limnology of lake torrens, an episodic salt lake of central Australia, with particular reference to unique events in 1989. *Hydrobiologia* **384**, 101–110.

Effects of Floods on Distribution and Reproduction of Aquatic Birds

ALDO POIANI

SUMMARY

Floods can have both beneficial and detrimental effects on aquatic bird reproduction and survival. Aquatic birds tend to disperse as an immediate response
to floods, but they will then gather on flooded areas where food sources are
abundant in order to breed. Large concentrations of birds tend to be prominent
as floodwaters recede and both adult and young birds concentrate at remaining
water-bodies. Increased local densities of breeding birds may also carry some
costs in terms of transmission of parasites and diseases. Adaptations of aquatic
birds to unpredictability of flooding are manifest in their ability to reproduce
opportunistically in response to changes in environmental conditions.

I. INTRODUCTION

Australia has over 150 species of waterbirds and shorebirds from the orders
Anseriformes, Podicipediformes, Pelecaniformes, Ciconiiformes, Gruiformes,
and Charadriiformes (Christidis and Boles, 1994), which inhabit freshwater

ADVANCES IN ECOLOGICAL RESEARCH VOL. 39
0065-2504/06 $35.00
DOI: 10.1016/S0065-2504(06)39004-6

bodies in coastal and inland regions. Aquatic birds (a general category that I will use to refer to waterbirds, including waterfowl, and shorebirds together) face an extremely variable landscape, both spatially and temporally, across the continent with climatic regions varying from an Equatorial and Tropical north to a Temperate southern region separated by a coastal Subtropical fringe and very substantial semiarid Grassland and arid Desert regions in the center (Stern *et al.*, 2000), the latter two accounting for 70% of continental Australia (Stafford-Smith and Morton, 1990). Availability of water and, therefore, good habitat for aquatic birds' survival and reproduction depends on frequency and magnitude of rainfall events and floods that occur mainly in autumn–winter (April to August) and spring (September and October) in the temperate south (e.g., parts of the Murray–Darling Basin), and in the wet season (November to April) in the tropical north that is under monsoonal influence.

Monsoonal rains in northern Australia are especially important, not only because some of the precipitation may eventually drain into the arid regions of central Australia, giving rise to large endorheic drainage basins (i.e., basins with an internal drainage system and therefore no water overflow) such as the Lake Eyre Basin (Fig. 1) (Roshier *et al.*, 2001a), but they can also affect southern, relatively more temperate regions through flooding of rivers such as the Darling in the Murray–Darling Basin (Fig. 1). How have aquatic birds adapted to the boom and bust regime of floodwater—and therefore food— availability across the continent? Do aquatic bird life histories reflect selection pressures imposed by an unpredictable environment? What landscape-level effects have floods on aquatic bird populations? How do birds "know" that a flood is coming? What are the conservation issues imposed by flood regulation and global climate phenomena?

The purpose of this chapter is to address the above-mentioned questions, narrow down some potential answers, and hopefully open up more questions to direct future research.

II. FLOODS AND AQUATIC BIRD LIFE HISTORIES

A. Survival

Most aquatic bird species live for more than 10 years in the wild (Klimkiewicz and Futcher, 1989). Therefore, they are likely to experience a flood event at least once in their lifetime, but many individual birds are likely to experience multiple floods. In an 11-year study covering the period from 1986 to 1997, Roshier *et al.* (2001b) showed that the Lake Eyre Basin was flooded four times, suggesting that adult aquatic birds may be able to use the same location for breeding, following flood events, more than once in their

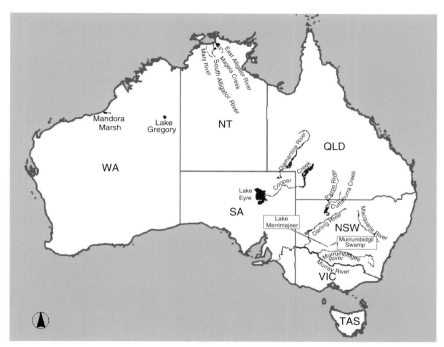

Plate 10

Plate 11

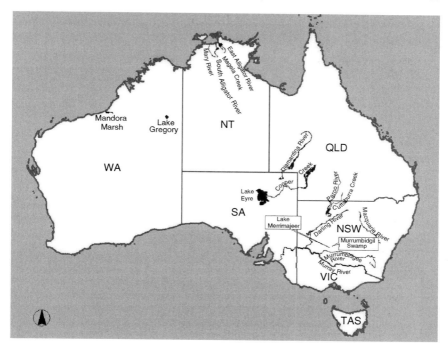

Figure 1 Map of Australia with the main water-bodies mentioned in the text. WA, Western Australia; NT, Northern Territory; SA, South Australia; QLD, Queensland; NSW, New South Wales; VIC, Victoria; TAS, Tasmania. (See Color Plate 10.)

lifetime. Even if we consider a 12.5-year period as the time interval between major flood events that fill Lake Eyre (Kingsford and Norman, 2002), this still is well within the average range of aquatic birds lifespan. Unfortunately, information on birds' use of flooded areas relies mainly on broad estimates of population sizes, whereas data on multiple sightings of banded individuals are scarce. However, it is well known that permanent water-bodies, even in the arid zone (e.g., in the Paroo catchment), can be refugia for long-living species, especially those that do not easily disperse over long distances. Those refugia may allow adults of some waterbird species (e.g., ducks, rallids) to survive until the next flood event (Roshier et al., 2001a).

Floodwaters carry food items for birds that may become concentrated in specific areas along the catchment. Moreover, water availability may induce reproductive events among fish and invertebrates in flooded areas that provide food sources for the survival and reproduction of aquatic birds (see also Section II.B) (Llewellyn, 1983; Maher and Carpenter, 1984; Crome, 1986; Tulloch et al., 1988; Briggs, 1990, 1992; Morton et al., 1993; Ausden et al., 2001; Kingsford and Norman, 2002).

Floods, however, may also carry risks for survival, with ground-nesting birds being potentially the most affected. Sometimes negative effects of floods on bird survival may even explain the decline of otherwise quite prolific species. Beauchamp (1997) suggested that the weka (*Gallirallus australis greyi*), a rallid that was studied in the North Island of New Zealand, had declined in the area in spite of being an omnivorous species capable of raising four broods per year in good environmental conditions. Beauchamp (1997) suggests that the decline may be partially explained by floods.

There are both direct and indirect risks for survival to aquatic birds during flooding. Other than drowning, there also may be increased transmission of waterborne pathogens and pathogens carried by mosquitoes. In southern Queensland, wet years are associated with decreased number of most water-bird species. One explanation may be the potential impact of botulism (*Clostridium botulinum*) on bird survival (Woodall, 1985). Other potential waterborne pathogens are *Giardia* spp., *Haemoproteus nettionis*, and *Plasmodium relictum* (Kingsford and Norman, 2002). Arboviruses (i.e., arthropod-transmitted viruses), such as flaviviruses (e.g., *West Nile virus*), also could negatively affect survival of susceptible aquatic birds during flood periods (Van der Meulen *et al.*, 2005).

An indirect risk of flooding to survival of aquatic birds is, paradoxically, the surge of reproductive activity triggered by floods, especially if reproduction occurs late in the season, followed by gradual drying of the flooded area. An increased level of bird mortality was detected at Cooper Creek and other flooded areas in the Lake Eyre Basin as the basin dried. Most of the mortality was accounted for by immature birds of that season (Kingsford *et al.*, 1999a). Whether a flood event could directly or indirectly jeopardize birds' survival, a significant proportion of adult birds will probably have the capability to disperse from the flooded area (Section II.D).

B. Reproduction

One of the most dramatic effects of floods on aquatic birds' life histories is increased reproduction (Fig. 2). Most aquatic birds in Australia breed opportunistically following floods. Whether this reproductive readiness is a result of a continuously activated neuroendocrine state sustaining gonadal development or a rapid switching of reproductive mode as a response to environmental cues associated with flooding is still uncertain (Deviche and Sharp, 2001). Species facing high environmental unpredictability are expected to adopt an opportunistic breeding strategy compounded by an increased tendency to nomadism (Hau *et al.*, 2004). Indeed, many of the aquatic bird species breeding in flooded areas also are nomadic (Section II.D).

Figure 2 Australian Pelican (*Pelecanus conspicillatus*) breeding colony at Lake Eyre. Photograph was taken in April 2000, after flooding of the lake. © Clive Minton. (See Color Plate 11.)

Breeding in flooded areas is associated with high local availability of food (Llewellyn, 1983; Milsom *et al.*, 2002; see also chapter by Sam Lake, Nick Bond, and Paul Reich). Dry–wet environmental cycles facilitate accumulation and release of nutrients through growth, reproduction, and decomposition of aquatic plants. Plants grow during the wet period of the cycle and die in the dry. Plant decomposition releases nutrients that are dissolved in water during the following wet period. Dissolved nutrients sustain population growth of invertebrates (e.g., dipteran larvae, crustaceans) that are staple food to many aquatic birds (Crome, 1986; Briggs, 1992). Increased local availability of prey also may be enhanced, in the short term, by floods as raising water levels flush out and concentrate macroinvertebrates, producing a pulse of food availability for bird breeding (Ausden *et al.*, 2001). The combined effect of local concentration and enhanced reproductive output following floods may also account for birds' vertebrate prey, such as fish, so that piscivorous birds also experience favorable conditions for breeding in flooded areas. When food availability is particularly enhanced following floods, as it occurs in central Australia, widespread reproductive events involving thousands of nesting birds may follow (Kingsford and Norman, 2002).

Larvae of chironomids (Diptera: Chironomidae) are important food for many Australian waterbirds. Maher and Carpenter (1984) carried out a study at Murrumbidgil swamp and Lake Merrimajeel in New South Wales and reported that populations of chironomids increased with rising water levels. Waterfowl breeding activity was positively correlated with average larvae density of *Chironomus tepperi*, which is known to colonize flooded areas. Increased water depth also was associated with increased breeding success of Australian white ibis (*Threskiornis molucca*) in a winter swamp near Ballarat (southeastern Australia) (Kentish, 1999). This effect might be due to either increased food availability, increased protection from terrestrial predators, or both. Floods provide good conditions for breeding along the rest of the Murray–Darling Basin as well. Of the 37 species of waterbirds reported by Briggs (1990) breeding in the Murray–Darling Basin, most do so after wetland flooding. Only two species were reported breeding seasonally irrespective of flood levels: the blue-billed duck (*Oxyura australis*) and the musk duck (*Biziura lobata*). The reason why these species are exceptions is not known but the fact that at least one of them (*B. lobata*) has young that, unique among ducks, are fed by the female parent (male breeders do not care for ducklings in either species) until they are almost fully grown (Marchant and Higgins, 1990) may be part of the explanation. Extended parental care provided to precocial young may increase probability of survival of ducklings, especially when environmental conditions are not optimal.

In southeast Queensland, McKilligan (2001) found that the number of nests of the cattle egret (*Ardea ibis*) was positively associated with amounts of local and regional rainfall, and so too was the mean size of advanced broods. In the wetlands of the Macquarie River in New South Wales, the number of waterbird nests was exponentially positively associated with annual water flow. The exponential effect (over the range of water flows studied) was mainly related to the breeding activity of one species: the straw-necked ibis (*T. spinicollis*) (Kingsford and Johnson, 1998). At Murrumbidgil swamp in western New South Wales, Crome (1986) also reported peak breeding in coincidence with spring flooding, and more nests were observed the drier the conditions were in the previous summer.

Although floods may provide favorable conditions for breeding through increased access to food resources, they may also carry a cost to breeding aquatic birds as unpredictably changing floodwater levels may inundate nests and kill eggs and/or nestlings. Ground-nesting altricial species are especially at risk (Smith and Renken, 1993; Sanders and Maloney, 2002; Gilbert and Servello, 2005). This may lead to a selective pressure on ground-nesting aquatic birds breeding in flood-prone areas to reduce the nesting period (e.g., through increased rate of development of eggs and nestlings) (Burger and Gochfeld, 1996). This is a hypothesis that could be tested through comparative analyses. However, even tree-nesting species may encounter problems with survival of offspring in flood-prone areas. A negative impact of recurrent floods on tree-nesting waterbirds is death of the trees due to waterlogging (Kozlowski, 2002). In fact Briggs and Thornton (1999) found that in the Murrumbidgee wetlands nesting activity increased with duration of floods, but it did so when water flooded *live* river red gum trees (*Eucalyptus camaldulensis*). In addition, Baxter (1994) found that if large nesting trees die as a consequence of prolonged flooding, egrets (*Ardea alba*, *Ardea intermedia*, *Ardea ibis*, and *Egretta garzetta*) abandon the site. Increased water levels that may decrease reproductive success also may be associated with increased nest predation and nest parasitism in some regions (Kruse *et al.*, 2003). Moreover, the usual association between floods and increased food availability should not be taken for granted. Following the 1993 record floods of the Mississippi River, clutch sizes laid by the great blue heron (*Ardea herodias*) were larger in less flooded areas than in more flooded areas, and birds also initiated nesting earlier in less flooded areas (Custer *et al.*, 1996). The effect was probably due to negative impact of record floods on food availability (Custer *et al.*, 1996).

Both positive and negative effects of floods on breeding success have been reported for magpie geese (*Anseranas semipalmata*). *Anseranas semipalmata* was studied in the floodplain of the Mary River, Northern Territory, a tropical region under monsoonal regime, by Whitehead and Tschirner (1990). The geese nest colonially and synchronously on flooded plains, however, because nests are built on platforms over water in emergent vegetation they may be lost especially after heavy floods (Whitehead and Tschirner, 1990). On the other hand if the wet season is delayed, nest densities decrease in the area (Whitehead and Saalfeld, 2000). This suggests an optimal level of flooding for reproduction: too much may be as bad as too little. Tulloch *et al.* (1988) reported a negative correlation between water depth and number of nests of magpie geese, but a positive correlation between number of nests and abundance of aquatic macrophytes that are used both as food and nest material by breeding geese in the floodplain of the Mary River.

In the Australian arid zone, most breeding of aquatic birds in flooded areas occurs several months after flooding (Maher and Braithwaite, 1992). Halse *et al.* (1998) reported data from a survey of aquatic birds carried out at Lake Gregory in northern Western Australia, an area under monsoonal influence, between 1989 and 1995. I analyzed Halse *et al.*'s (1998) data appearing in their Tables 2 and 3 excluding March data, to control for seasonal effects, and also data from June 1991 to avoid pseudoreplication. My analyses indicate that both number of breeding species (Pearson product–moment correlation: $r_3 = 0.81$, $P < 0.05$) and number of nests or broods ($r_3 = 0.94$, $P < 0.02$) were positively correlated with water depth (see also Kingsford and Auld, 2005). No significant correlation was detected between water depth and total number of species ($r_3 = 0.18$, $P = 0.31$) or population densities ($r_3 = 0.01$, $P > 0.40$). Abundance and nesting were not correlated either ($r_3 = 0.08$, $P > 0.25$) (Table 1). Both population density

Table 1 Water depth and some life-history variables of aquatic birds at Lake Gregory

Date[a]	Water depth (m)	Number of species	Population size estimates	Number of breeding species	Number of nests or broods
October 1989	1.5	50	81,618	1	6
September 1991	7.4	46	133,227	11	820
August 1993	11.7	47	71,940	14	1444
May 1994	10.8	44	80,000	8	1915
October 1995	9.3	57	200,000	7	440

[a]Data are from Tables 2 and 3 of Halse *et al.* (1998).

and number of nests or broods values were log-transformed before entering them in the analyses. The results suggest that floods in the arid zone attract two categories of aquatic birds: one that transfer to the flooded area to, presumably, mainly improve body condition (e.g., some waders and some rallids), whereas others also transfer to breed (e.g., black swan *Cygnus atratus*, cormorants, Eurasian coot *Fulica atra*, Caspian tern *Sterna caspia*).

In a study carried out by Reid (2004) at Cooper Creek and Diamantina River and associated wetlands that are part of the Lake Eyre Basin, abundance and nesting were also not correlated: Pearson product–moment correlation $r_9 = 0.29$, $P > 0.10$ (Fig. 3). Reid's (2004) original data were reanalyzed by using genera as data points to, at least partially, control for phylogenetic effects (Harvey and Pagel, 1991). Values for each variable for the genus are the mean value for that variable for congeneric species and were entered in the correlation analysis after log-transformation. As for Lake Gregory, this result for central Australia again suggests that flooded, inland basins may exert a combined effect on aquatic birds as refugia for survival (e.g., *Malacorhynchus* and *Aythya*), breeding grounds (e.g., *Phalacrocorax* and *Egretta*), or both (e.g., *Pelecanus*). Halse *et al.* (2005) recorded 61 species of aquatic birds in the arid

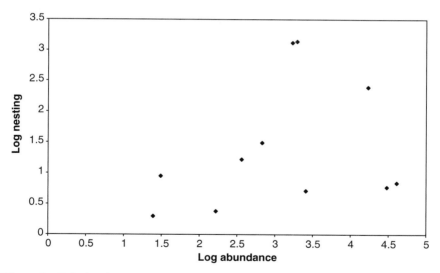

Figure 3 Relation between abundance and nesting of waterbirds in the Lake Eyre Basin. Raw data to calculate values for each genus were taken from Reid (2004) and are from surveys carried out in May 2004. Genera in ascending order of log abundance are: *Nycticorax*, *Anhinga*, *Ardea*, *Platalea*, *Threskiornis*, *Egretta*, *Phalacrocorax*, *Anas*, *Pelecanus*, *Aythya*, and *Malacorhynchus*.

zone Mandora Marsh, northwestern Australia, of these species, 44.3% were recorded breeding in the marsh.

C. Population Density

The general pattern of temporal fluctuations in population densities of aquatic birds in relation to floods is that low numbers usually occur soon after flooding, then population sizes build in the flooded areas as birds feed and then breed there, with largest concentrations of birds being detected during dry periods, when birds converge at remnant water-bodies (Kingsford and Norman, 2002). The low number of aquatic birds usually counted during or soon after flooding events may in part be due to local movement of birds away from environmental stress (Morant and O'Callaghan, 1990). In a study of waterbird populations in eastern Australia, Kingsford et al. (1999b) found that population sizes of 16 species were negatively correlated with the amount of rain in the study area and positively correlated with the amount of rainfall outside the study area, suggesting that rain triggers movement away from areas being flooded. The above pattern, however, is not general for all guilds of aquatic birds. For instance, although Roshier et al. (2001a) reported an immediate negative effect of floods on numbers of dabbling ducks, the reverse was true for fish-eating species, deepwater foragers, and diving ducks all of which increased in numbers following floods in the Lake Eyre Basin. Local abundance of fish being transported with floodwaters may explain the pattern. Morton et al. (1993) studied aquatic birds' abundance and distribution in five floodplains in the Alligator River region, Northern Territory. I carried out a Wilcoxon two-sample test on the population sizes data available in their Table 7: monthly average counts in the wet season (November to April) versus counts in the dry season (May to October) for Anhinga melanogaster, Phalacrocorax sulcirostris, Phalacrocorax melanoleucos, Sterna nilotica, Sterna caspia, and Porphyrio porphyrio in each of the five floodplains. Across species and floodplains there are significant differences in population sizes between seasons ($Z = 3.17$, $P < 0.001$) with trends indicating that those waterbirds are more abundant in the area during the dry season, with extreme examples as in the case of Porphyrio porphyrio at Boggy Plain. Exceptions to the trend are Phalacrocorax sulcirostris, Phalacrocorax melanoleucos, and Sterna caspia at Cooper Creek, perhaps indicating increased availability of fish in the area during the wet season. Population densities of piscivorous waterbirds followed the same trend also in southern Queensland following floods. At two localities, one coastal (Moreton Bay) and one inland (Warrego), Woodall (1985) showed that waterbird numbers decreased in both areas with increasing rainfall, with the exception of two piscivorous pelecaniforms: Phalacrocorax varius at Warrego and Anhinga melanogaster at Moreton Bay.

At the Paroo River and Cuttaburra Creek in New South Wales, Maher and Braithwaite (1992) reported maximum density of aquatic birds (mainly Anatidae, 80.3%) 28.5 months after flooding, with density peaks differing among species. A *leitmotif* of population dynamics of waterbirds at the site is that the highest density coincided with the drying phase of the wetland. The pattern is also expected from recruitment of new individuals after breeding, which mostly occurred 4 and 9 months following the floods. *Irediparra gallinacea* jacanas showed the same population dynamics in the floodplains of the Magela Creek, a tributary of the East Alligator River: low population densities during the wet season and increased densities during the dry season, mainly resulting from dispersal movement during the wet as water becomes available across the landscape and concentration at permanent water pools during the dry (Dostine and Morton, 2000). A trend for increased population sizes of waterbirds in wetlands during the dry season is not a pattern unique to Australian birds (e.g., Herremans, 1999).

At Lake Eyre total abundances of waterbirds and shorebirds concentrating in the area in response to major flood events may be relatively consistent between floods (see Table 2, data are from Lane, 1984, 1987 and Kingsford and Porter, 1993) (Wilcoxon two-sample test, $P > 0.50$, two-tailed). However, patterns for individual species can vary dramatically with small shorebirds (Charadriiformes) and gray teals (*Anas gracilis*) increasing by an order of magnitude, while *Pelecanus conspicillatus*, *Phalacrocorax varius*, and large shorebirds decreased in numbers between flood years (Kingsford and Porter, 1993).

Both total water flow levels and the shape of the flooded lake, its quality as a habitat for aquatic birds, and its isolation from other such water-bodies can be important factors affecting distribution and abundance of aquatic birds. Kingsford and Johnson (1998) studied the effect of water flow on abundance

Table 2 Aquatic bird counts at Lake Eyre North

Species	1984[a]	1990–1991[b]
Pelecanus conspicillatus	17,389	4648
Phalacrocorax carbo	20	47
Phalacrocorax varius	610	20
Anas gracilis	682	61,918
Himantopus himantopus	16	150
Cladorhynchus leucocephalus	23,910	66,206
Recurvirostra novaehollandiae	95,600	30,714
Small shorebirds	15,864	136,768

[a]Data are from Lane (1984, 1987).
[b]Data are from Kingsford and Porter (1993).

of waterbirds at the Macquarie Marshes. Over a 12-year period, colony sizes of all six species of waterbird studied (*Ardea intermedia*, *Nycticorax caledonicus*, *Plegadis falcinellus*, *Threskiornis spinicollis*, *Threskiornis molucca*, and *Platalea regia*) were positively correlated with total annual flow. Consistent with the above result, when levels of flooding declined, total population densities and species richness of waterbirds also decreased (Kingsford and Thomas, 1995). A survey of 28 wetlands and 43 waterbird species carried out in New South Wales indicate that mean values of abundance of all waterbird species increases with larger size of natural wetlands (Roshier *et al.*, 2002). Kingsford *et al.* (1999b) also detected a positive correlation between population sizes of waterbirds and wetland area in eastern Australia. However, more complex patterns emerge among different categories of birds. For example, piscivores, some grazing waterfowl, shoreline foragers, and large waders are more abundant in larger wetlands, while small waders and dabbling ducks are rarer in larger wetlands (Roshier *et al.*, 2002).

The usually positive effect of size of the flooded area on aquatic bird population sizes may be affected by landscape-scale effects. This general phenomenon can be illustrated by studies carried out both in and outside Australia. The number of Podicipediformes, Anseriformes, and Rallidae species surveyed by Paracuellos and Tellería (2004) in the Almería region of southeastern Spain in a system of 26 wetland ponds were positively correlated with pond area and negatively correlated with pond isolation in winter, while in spring–summer pond area and shore coverage by emergent vegetation are the best predictors, being positively correlated with number of species. Herremans (1999) also suggested that as many waterbirds use shore or nearshore habitats for feeding and breeding, the shape of the lake or flooded area (perimeter to area ratio) should be more relevant than size. Wetland quality as such may stand out above all other variables as predictor of aquatic bird population sizes. In their review study of Australian wetlands, Roshier *et al.* (2001a) concluded that wetland area was a poor predictor of waterbird abundance. Moreover, counts of waterbirds were not correlated in adjacent wetlands, suggesting that wetland quality may be variable and determinant of population sizes of waterbirds.

D. Movements

Flooding provides both an environmental cue that triggers aquatic bird dispersal and also a process that increases landscape connectivity facilitating aquatic bird movements across the landscape. Kingsford *et al.* (1999b) suggested that onset of flooding may initially trigger movements of waterbirds away from the flooded area. In the Northern Territory, magpie geese disperse over floodplains during the wet season to nest, and contract their distributions as they concentrate in large numbers at permanent freshwater bodies during the dry season (Bayliss and Yeomans, 1990). In the Murray–Darling Basin, cormorants (Phalacrocoracidae) disperse along watercourses (Llewellyn, 1983) that may be modified by floods.

Ducks in the arid region of Australia usually are nomadic, although some species have a strategy of settling on permanent water-bodies to breed during drought years and dispersing into flooded areas following heavy rains (Briggs, 1992). Large movements of birds across the continent following floods are best exemplified by the large concentration of aquatic birds in flooded endorheic lakes. For instance, most waterbirds that concentrate in the Lake Eyre Basin following floods come from outside the basin (Reid, 2004). Wetland connectivity may be increased after flooding leading to increased chances of successful, long-distance dispersal events (Kingsford and Norman, 2002) (Section IV).

By increasing connectivity of the landscape for waterbirds, floods promote movements across the landscape (Roshier et al., 2001a). The more capable a species is of moving over long distances, the more likely that species can persist in arid zones where water availability is unpredictable and permanent water-bodies are spatially scattered. Species with less ability to disperse either become locally extinct or are "trapped" in isolated, permanent water-bodies.

Roshier et al. (2001a) proposed the hypothesis that movements of waterbirds from arid to mesic zones of the Australian continent are an effect of a gradual increase in aridity in the center, rather than an outcome of movements toward the arid zone in response to flooding. Both possibilities may be correct as they can be valid answers to problems at two different levels of analysis. The long-term trend toward a reduced residence of aquatic bird populations in the Australian arid zone may be a result of climatic changes that led to increased aridity. However, short-term response of aquatic birds to floods in the arid zone (e.g., long-distance movements from the coast to the center) may be a behavioral pattern maintained by current adaptive fitness benefits of reproducing in a predator-safe and—for a short period of time—also food-rich environment.

III. HOW DO BIRDS KNOW IT IS FLOODING?

For aquatic birds that are resident in the area subject to a flood, the answer to the above question is simple and straightforward: They know it is flooding because they are getting wet! But, what about birds that live far away, sometimes hundreds of kilometers, from the flood area and yet fly to the newly formed lakes and swamps to feed and in some cases breed? How do they know that at the end of their journey they are going to land on a food-rich floodplain?

In order to reach their destination (e.g., a flooded endorheic lake in the arid zone) that can be hundreds of kilometers away, aquatic birds have to make use of both a map sense (i.e., an ability to determine their position relative to destination) and a compass sense (i.e., an ability to maintain that direction in flight). Experimental evidence is available from several species that suggest that birds can navigate over long distances by using gradients of trace gases, direction and magnitude of the geomagnetic field, the Sun and other stars, infrasounds, and topography (Phillips, 1996; Hagstrum, 2000; Wallraff and Andreae, 2000; Wallraff, 2003).

Some species can derive map information by using the magnetic receptors associated with their trigeminal nerve (Phillips, 1996) that are sensitive to changes in direction and intensity of the geomagnetic field. Other species can home using cues from variable distances of 100–700 km (e.g., Wallraff, 2003). For instance, Wallraff and Andreae (2000) showed that trace gases, such as volatile organic compounds, have predictable spatial gradients that may be used by birds for navigation. Hagstrum (2000) provided circumstantial evidence that atmospheric infrasounds can also be a source of directional information to birds. Infrasounds can be detected hundreds to thousands of kilometers from the source of emission and they can be produced by thunderstorms and other atmospheric phenomena. Topographic cues and, in general, visual landscape features also can be used by birds to navigate (Hagstrum, 2000; Wallraff, 2003), and so it can the position of the Sun and other stars for some species (Hagstrum, 2000).

To the best of my knowledge, there are no published studies on Australian aquatic birds that investigate the mechanisms of navigation used by species in response to a flood occurring hundreds of kilometers away. However, I may venture the following potential scenario to perhaps narrow down a promising area for future research. I already mentioned in Section II.A that most aquatic birds are relatively long-living species that are likely to individually experience more than one flood event in their lifetime. Of course, inexperienced birds may always learn their way to the newly formed waterbody by following more experienced individuals. But, what kind of clues are experienced birds using in order to start and direct their flight and also to navigate? What is telling them that an area far away is flooded and how do they get there? Are all birds (both young and old) endowed with hard-wired sensory mechanisms that trigger onset and maintain direction of flight? If birds possess the ability to detect changes in atmospheric infrasounds caused by thunderstorms and integrate those changes in their memory over a period of time, then they may use those faculties to perceive that floods may be occurring in the general direction of the source of the infrasounds. Birds may subsequently use gradients of volatile chemicals that originate in the flooded area as a compass. Subsequent journeys may be made more efficient by memorizing topographic cues and changes in the geomagnetic field. Potential use of the Sun and/or other stars to maintain their flight direction is also a possibility, especially for long-distance migratory species such as waders. The use of several cues may make navigation more precise and more energy efficient. Which senses and environmental cues birds use either simultaneously or in sequence is an area of future research that holds the promise of exciting discoveries.

IV. THE LANDSCAPE DIMENSION

Abundance, diversity, and distribution of aquatic birds are highly dependent on availability of appropriate habitat at both local and regional scales, depending on the ability of bird species to move across the landscape. Roshier et al. (2001c) suggested that floods increase both habitat availability and habitat connectivity, facilitating movements of waterbirds. In the Australian arid region, habitats sustaining resident breeding populations of aquatic birds may become available if floods are frequent enough to maintain permanent water-bodies at an appropriate distance from each other for birds to disperse (Roshier et al., 2001a). Increased connectivity of wetlands also can be of great conservation value for intercontinental migratory shorebirds (Haig et al., 1998).

Table 3 summarizes population sizes of shorebirds recorded at six lakes in the arid and Grassland zones of Australia, area of the lake and recurrence of inundation (data are taken from Appendix 1 of Roshier et al., 2001b and Watkins, 1993). A multiple correlation analysis (after log-transformation of population size and maximum area of the lake) suggests that population size of shorebirds increases with the area of the lake ($r_4 = 0.75$, $P = 0.04$) but it is not affected by recurrence of inundation ($r_4 = -0.08$, $P = 0.43$). However, the positive relationship between population size and maximum area becomes marginally nonsignificant after controlling for the effect of frequency of inundation (partial-$r_3 = 0.79$, $P = 0.056$).

Table 3 Population sizes of shorebirds, maximum lake area, and recurrence of inundation at six lakes in the Australian Arid and Grassland zones

Lake[a]	Number of individuals	Maximum area (ha)	Recurrence of inundation (%)
Cowal	3500	4140	88.9
Numalla	2500	2953	86.1
Gregory	20,000	24,881	76.7
Eyre	129,000	868,975	62.9
Buloke	16,880	5515	63.5
Goran	300	6381	58.3

[a]Data are from Roshier et al. (2001b) and Watkins (1993).

V. CONSERVATION ISSUES

As I mentioned in Sections II.A and B, floods sometimes may be a threat to survival and reproductive success of aquatic birds (Beauchamp, 1997). In the case of ground-nesting birds, the most obvious target for conservation efforts in the case of floods, the provision of elevated platforms to build the nest may be enough to overcome the effect of raising water levels. However, providing elevated platforms for nesting is a strategy that does not always pay off (Rounds et al., 2004).

Floods, however, also represent an opportunity for breeding for many aquatic birds, while at the same time being perceived as a threat to life and economic productivity by human populations resident in flood-prone areas (see chapters by John Freebairn and by Catherine Allan, Allan Curtis, and Nicki Mazur). The latter has led to many instances of water-flow regulation and flood-management strategies that may impinge on long-term viability of local populations of aquatic birds. Flood regulation may alter the natural pulses of river discharges (the Flood Pulse Concept of Junk et al., 1989) that affect predictability of successful reproduction by aquatic birds (Puckridge et al., 1998).

The Murray–Darling Basin is a very good case study for the effects of flood regulation on viability of aquatic bird populations. Wetlands in the Murray–Darling Basin do not fill and dry regularly following natural floods, but water levels are dictated by irrigation needs and flood-regulation strategies (Briggs et al., 1997). These alterations of the natural flood patterns along the Murray–Darling Basin have been associated with some negative effects on waterbird reproduction (Kingsford and Norman, 2002).

In a 19-year-long study of six floodplain systems in southeastern Australia, Kingsford et al. (2004) found that, at least in the month of October, unregulated lakes had a higher mean density of waterbirds than regulated lakes, with Anseriformes and herbivorous species being relatively more abundant in unregulated lakes, while piscivorous species prevailed in regulated lakes. Total species richness also was greater in unregulated lakes. Regulated lakes usually have less shallow-water areas that prevent both the establishment of aquatic macrophytes used by waterfowl and small waders. Olsen and Weston (2004) convene on issuing the same warning regarding potential effects of flood regulation on waterbird conservation. In their report on the state of Australian birds in wetlands, Olsen and Weston (2004) argued that flood mitigation is a significant threat to wetlands and, therefore, waterbird conservation. In their comparison of data from the Atlas of Australian Birds 1977–1981 and 1998–2002, they concluded that of 36 waterbird species 6 (16.6%) have increased in numbers, 22 (61.1%) have not changed, and 8 (22.2%) have declined. Detailed studies of which habitat changes are induced by flood regulation and in which way those changes alter survival and reproduction of the different aquatic bird species will help design better strategies for the use of freshwater for both human economic activity and environmental conservation purposes.

How are global climatic phenomena, such as global warming and the El Niño–Southern Oscillation (ENSO), going to affect future viability of Australian aquatic bird populations? Global climatic change models generally predict drier summers and wetter winter–springs in temperate regions of the planet (Sorenson *et al.*, 1998; Watkinson *et al.*, 2004), with differing effects on aquatic bird survival and reproduction. Increased wet conditions may lead to higher reproductive success while extended drought periods may result in decreased breeding and survival of aquatic birds. Moreover, elevated population densities at wetlands during a drought also may increase outbreak of diseases with negative consequences for both birds and humans. Predictions for Australian aquatic bird populations depend on the projected climatic trends for the tropical north and the temperate south. The southern part of the continent is expected to generally dry (Chambers *et al.*, 2005) with potential negative effects on viability of aquatic bird populations due to loss of habitat. However, potential increase in precipitation in the tropical north is likely to be beneficial to populations of aquatic birds. It is predicted that an increase of about 10% in mean annual rainfall in tropical Australia may be enough to transform endorheic lakes, such as Lake Eyre, into permanent water-bodies that can sustain large populations of aquatic birds over a longer period than currently (Roshier *et al.*, 2001a,b).

In addition to potential effects of global warming on aquatic bird populations, Australia also is significantly affected by ENSO. Eastern Australia is subject to high intensity of flooding during La Niña, the opposite phase of ENSO. Norman and Nicholls (1991) (see also Kingsford *et al.*, 1999b) reported a negative correlation (with a 24-month lag) between numbers of waterfowl in hunters' bags in Victoria (southeastern Australia) and values of the Southern Oscillation Index, which is the normalized atmospheric pressure difference between Papeete and Darwin. Norman and Nicholls (1991) related this effect to increased reproductive rates of waterfowl during the wet phase and high concentration of birds at remaining water-bodies during the dry ENSO years.

VI. CONCLUSIONS

Floods are unpredictable environmental perturbations that have both short- and long-term effects on populations of aquatic birds. Floodwater may transport nutrients but may destroy ground-nesting sites or nesting trees. How have aquatic birds adapted to the boom and bust regime of floods? Birds can move away from flooded areas when a flood is perceived as a danger to survival, but they also can track availability of flooded habitats, even from long distances, to use them for breeding when local conditions allow. This ability to exploit a temporally and spatially unpredictable resource is facilitated by birds' reproductive readiness and still unknown mechanisms of navigation. By increasing landscape connectivity for aquatic birds, floods also allow dispersal and facilitate survival of populations that might become trapped in isolated water-bodies. Population viability of aquatic birds may be negatively affected by the compounded effects of global climatic phenomena and anthropogenic activities such as flood regulation. Such effects are most likely to negatively impact on wild aquatic bird populations in the southern parts of Australia.

ACKNOWLEDGMENTS

I am grateful to Clive Minton for providing the photograph for Fig. 2, and to Sue Argus for drawing Fig. 1. David Morgan (Department of Zoology, University of Melbourne) and the Department of Botany, La Trobe University, kindly allowed me the use of their computer facilities. I am also grateful to Ralph Mac Nally and Hugh Ford for their constructive comments on the chapter.

REFERENCES

Ausden, M., Sutherland, W.J. and James, R. (2001) The effects of flooding lowland wet grassland soil macroinvertebrate prey of breeding wading birds. *J. Appl. Ecol.* **38**, 320–338.

Baxter, G.S. (1994) The location and status of egret colonies in coastal New South Wales. *Emu* **94**, 255–262.

Bayliss, P. and Yeomans, K.M. (1990) Seasonal distribution and abundance of magpie geese, *Anseranas semipalmata* Latham, in the Northern Territory, and their relationship to habitat, 1983–86. *Aust. Wildl. Res.* **17**, 15–38.

Beauchamp, A.J. (1997) The decline of the North Island Weka (*Gallirallus australis greyi*) in the East Cape and Opotiki Regions, North Island New Zealand. *Notornis* **44**, 27–35.

Briggs, S. (1990) Waterbirds. In: *The Murray* (Ed. by N. Mackay and D. Eastburn), pp. 336–344. Murray Darling Basin Commission, Canberra.

Briggs, S. (1992) Movement patterns and breeding characteristics of arid zone ducks. *Corella* **16**, 15–22.

Briggs, S.V. and Thornton, S.A. (1999) Management of water regimes in River Red Gum *Eucalyptus camaldulensis* wetlands for waterbird breeding. *Aust. Zool.* **31**, 187–197.

Briggs, S.V., Thornton, S.A. and Lawler, W.G. (1997) Relationships between hydro-logical control of river red gum wetlands and waterbird breeding. *Emu* **97**, 31–42.

Burger, J. and Gochfeld, M. (1996) Use of space by nesting black-billed gulls *Larus bulleri*: Behavioural changes during the reproductive cycle. *Emu* **96**, 73–80.

Chambers, L.E., Hughes, L. and Weston, M.A. (2005) Climate change and its impact on Australia's avifauna. *Emu* **105**, 1–20.

Christidis, L. and Boles, W.E. (1994) *The Taxonomy and Species of Birds of Australia and Its Territories*. Royal Australasian Ornithologists Union, Hawthorn East.

Crome, F.H.J. (1986) Australian waterfowl do not necessarily breed on a rising water level. *Aust. Wildl. Res.* **13**, 461–480.

Custer, T.W., Hines, R.K. and Custer, C.M. (1996) Nest initiation and clutch size of great blue herons on the Mississippi river in relation to the 1993 flood. *Condor* **98**, 181–188.

Deviche, P. and Sharp, P.J. (2001) Reproductive endocrinology of a free-living, opportunistically breeding passerine (White-winged Crossbill, *Loxia leucoptera*). *Gen. Comp. Endocrinol.* **123**, 268–279.

Dostine, P.L. and Morton, S.R. (2000) Seasonal abundance and diet of the comb-crested jacana *Irediparra gallinacea* in the tropical Northern Territory. *Emu* **100**, 299–311.

Gilbert, A.T. and Servello, F.A. (2005) Water level dynamics in wetlands and nesting success of Black Terns in Maine. *Waterbirds* **28**, 181–187.

Hagstrum, J.T. (2000) Infrasound and the avian navigational map. *J. Exp. Biol.* **203**, 1103–1111.

Haig, S.M., Mehlman, D.W. and Oring, L.W. (1998) Avian movements and wetland connectivity in landscape conservation. *Conserv. Biol.* **12**, 749–758.

Halse, S.A., Pearson, G.B. and Kay, W.R. (1998) Arid zone networks in time and space: Waterbird use of Lake Gregory in North-Western Australia. *Int. J. Ecol. Environ. Sci.* **24**, 207–222.

Halse, S.A., Pearson, G.B., Hassell, C., Collins, P., Scanlon, M.D. and Minton, C.D.T. (2005) Mandora Marsh, north-western Australia an arid-zone wetland maintaining continental populations of waterbirds. *Emu* **105**, 115–125.

Harvey, P.H. and Pagel, M.D. (1991) *The Comparative Method in Evolutionary Biology*. Oxford University Press, Oxford.

Hau, M., Wikelski, M., Gwinner, H. and Gwinner, E. (2004) Timing of reproduction in a Darwin's finch: Temporal opportunism under spatial constraints. *Oikos* **106**, 489–500.

Herremans, M. (1999) Waterbird diversity, densities, communities and seasonality in the Kalahari Basin, Botswana. *J. Arid Environ.* **43**, 319–350.

Junk, W.J., Bayley, P.B. and Sparks, R.E. (1989) The flood pulse concept in river-floodplain systems. *Can. Spec. Publ. Fish. Aquat. Sci.* **106**, 110–127.

Kentish, B. (1999) Breeding of Australian White Ibis, Straw-necked Ibis and Silver Gulls on winter swamp, Ballarat. *Corella* **23**, 37–42.

Kingsford, R.T. and Auld, K.M. (2005) Waterbird breeding and environmental flow management in the Macquarie Marshes, Arid Australia. *River Res. Appl.* **21**, 187–200.

Kingsford, R.T. and Johnson, W. (1998) Impact of water diversions on colonially nesting waterbirds in the Macquarie marshes of arid Australia. *Colonial Water-birds* **21**, 159–170.

Kingsford, R.T. and Norman, F.I. (2002) Australian waterbirds—products of the continent's ecology. *Emu* **102**, 47–69.

Kingsford, R.T. and Porter, J.L. (1993) Waterbirds of Lake Eyre, Australia. *Biol. Conserv.* **65**, 141–151.

Kingsford, R.T. and Thomas, R.F. (1995) The Macquarie Marshes in arid Australia and their waterbirds: A 50-year history of decline. *Environ. Manag.* **19**, 867–878.

Kingsford, R.T., Curtin, A.L. and Porter, J. (1999a) Water flows on Cooper Creek in arid Australia determine 'boom' and 'bust' periods for waterbirds. *Biol. Conserv.* **88**, 231–248.

Kingsford, R.T., Wong, P.S., Braithwaite, L.W. and Maher, M.T. (1999b) Waterbird abundance in eastern Australia. *Wildl. Res.* **26**, 351–366.

Kingsford, R.T., Jenkins, K.M. and Porter, J.L. (2004) Imposed hydrological stability on lakes in arid Australia and effects on waterbirds. *Ecology* **85**, 2478–2492.

Klimkiewicz, M.K. and Futcher, A.G. (1989) Longevity records of North American birds. Supplement I. *J. Field Ornithol.* **60**, 469–494.

Kozlowski, T.T. (2002) Physiological-ecological impacts of flooding on riparian forest ecosystems. *Wetlands* **22**, 550–561.

Kruse, K.L., Lovvorn, J.R., Takekawa, J.Y. and Mackay, J. (2003) Long-term productivity of canvasbacks (*Aythya valisineria*) in a snowpack-driven desert marsh. *Auk* **120**, 107–119.

Lane, B. (1984) "Report on a Trip to Lake Eyre North 20–24 September 1984." RAOU, Water Studies Program, Moonee Ponds.

Lane, B. (1987) *Shorebirds in Australia*. Nelson Publishers, Melbourne.

Llewellyn, L.C. (1983) Movements of cormorants in south-eastern Australia and the influence of floods on breeding. *Aust. Wildl. Res.* **10**, 149–167.

McKilligan, N. (2001) Population dynamics of the Cattle Egret (*Ardea ibis*) in south-east Queensland: A 20-year study. *Emu* **101**, 1–5.

Maher, M. and Carpenter, S.M. (1984) Benthic studies of waterfowl breeding habitat in south-western New South Wales. II. Chironomid populations. *Aust. J. Mar. Freshwater Res.* **35**, 97–110.

Maher, M.T. and Braithwaite, L.W. (1992) Patterns of waterbird use in wetlands of the Paroo, a river system of inland Australia. *Rangeland J.* **14**, 128–142.

Marchant, S. and Higgins, P.J. (1990) *Handbook of Australian, New Zealand and Antarctic Birds. Volume 1 Ratites to Ducks*. Oxford University Press, Melbourne.

Milsom, T.P., Hart, J.D., Parkin, W.K. and Peel, S. (2002) Management of coastal grazing marshes for breeding waders: The importance of surface topography and wetness. *Biol. Conserv.* **103**, 199–207.

Morant, P.D. and O'Callaghan, M. (1990) Some observations on the impact of the March 1988 flood on the biota of the Orange River mouth. *T. Roy. Soc. S. Afr.* **47**, 295–305.

Morton, S.R., Brennan, K.G. and Armstrong, M.D. (1993) Distribution and abundance of grebes, pelicans, darters, cormorants, rails and terns in the Alligator Rivers region, Northern Territory. *Wildl. Res.* **20**, 203–217.

Norman, F.I. and Nicholls, N. (1991) The Southern Oscillation and variation in waterfowl abundance in southeastern Australia. *Aust. J. Ecol.* **16**, 485–490.

Olsen, P. and Weston, M. (2004) *The State of Australia's Birds 2004: Water, Wetlands and Birds*. RAOU, Hawthorn East.

Paracuellos, M. and Tellería, J.L. (2004) Factors affecting the distribution of a waterbird community: The role of habitat configuration and bird abundance. *Waterbirds* **27**, 446–453.

Phillips, J.B. (1996) Magnetic navigation. *J. Theor. Biol.* **180**, 309–319.

Puckridge, J.T., Sheldon, F., Walker, K.F. and Boulton, A.J. (1998) Flow variability and the ecology of large rivers. *Mar. Freshwat. Res.* **49**, 55–72.

Reid, J.R.W. (2004) "Aerial Survey of Waterbird Abundance and Breeding in the Coongie Lakes Wetlands and Goyder Lagoon, Lake Eyre Basin, SA, May 2004 General Report." Australian Government Department of Environment and Heritage, Canberra.

Roshier, D., Robertson, A. and Kingsford, R. (2001a) "The Availability of Wetland Habitat for Waterbirds in Arid Australia. Final Report to the National Wetlands research and Development Program Administered by Environment Australia." Charles Sturt University and National Parks and Wildlife Service, New South Wales.

Roshier, D.A., Whetton, O.H., Allan, R.J. and Robertson, A.I. (2001b) Distribution and persistence of temporary wetland habitats in arid Australia in relation to climate. *Aust. Ecol.* **26**, 371–384.

Roshier, D.A., Robertson, A.I., Kingsford, R.T. and Green, D.G. (2001c) Continental-scale interactions with temporary resources may explain the paradox of large populations of desert waterbirds in Australia. *Landsc. Ecol.* **16**, 547–556.

Roshier, D.A., Robertson, A.I. and Kingsford, R.T. (2002) Responses of waterbirds to flooding in an arid region of Australia and implications for conservation. *Biol. Conserv.* **106**, 399–411.

Rounds, R.A., Erwin, R.M. and Porter, J.H. (2004) Nest-site selection and hatching success of waterbirds in coastal Virginia: Some results of habitat manipulation. *J. Field Ornithol.* **75**, 317–424.

Sanders, M.D. and Maloney, R.F. (2002) Causes of mortality at nests of ground-nesting birds in the Upper Waitaki Basin, South Island, New Zealand: A 5-year video study. *Biol. Conserv.* **106**, 225–236.

Smith, J.W. and Renken, R.B. (1993) Reproductive success of Least terns in the Mississippi River Valley. *Colonial Waterbirds* **16**, 39–44.

Sorenson, L.G., Goldberg, R., Root, T.L. and Anderson, M.G. (1998) Potential effects of global warming on waterfowl populations breeding in the northern great plains. *Clim. Change* **40**, 343–369.

Stafford-Smith, D.M. and Morton, S.R. (1990) A framework for the ecology of arid Australia. *J. Arid Environ.* **18**, 255–278.

Stern, H., de Hoedt, G. and Ernst, J. (2000) Objective classification of Australian climates. *Aus. Met. Mag.* **49**, 87–96.

Tulloch, D., Cellier, K.M. and Hertog, A.L. (1988) The distribution of the nests of the Magpie Goose (*Anseranas semipalmata* Latham) at Kapalga, N.T.: A four-year study. *Aust. Wildl. Res.* **15**, 211–221.

Van der Meulen, K.M., Pensaert, M.B. and Nauwynck, H.J. (2005) West Nile virus in the vertebrate world. *Arch. Virol.* **150**, 637–657.

Wallraff, H.G. (2003) Zur olfaktorischen navigation der vögel. *J. Ornithol.* **144**, 1–32.

Wallraff, H.G. and Andreae, M.O. (2000) Spatial gradients in ratios of atmospheric trace gases: A study stimulated by experiments on bird navigation. *Tellus B* **52**, 1138–1157.

Watkins, D. (1993) "A National Plan for Shorebird Conservation in Australia." Australasian Wader Studies Group Royal Australasian Ornithologists Union and World Wide Fund for Nature. RAOU Report No. 90, Moonee Ponds.

Watkinson, A.R., Gill, J.A. and Hume, M. (2004) Flying in the face of climate change: A review of climate change, past, present and future. *Ibis* **146**, 4–10.

Whitehead, P.J. and Tschirner, K. (1990) Magpie goose, *Anseranas semipalmata*, nesting on the Mary River floodplain, Northern Territory, Australia: Extent and frequency of flooding losses. *Aust. Wildl. Res.* **17**, 147–157.

Whitehead, P.J. and Saalfeld, K. (2000) Nesting phenology of magpie goose (*Anseranas semipalmata*) in monsoonal northern Australia: Responses to antecedent rainfall. *J. Zool. (Lond.)* **251**, 495–508.

Woodall, P.F. (1985) Waterbird populations in the Brisbane Region, 1972–83, and correlates with rainfall and water heights. *Aust. Wildl. Res.* **12**, 495–506.

The Landscape Context of Flooding in the Murray–Darling Basin

ANDREA BALLINGER AND RALPH MAC NALLY

SUMMARY

Here we apply theoretical frameworks to understanding the influence of flooding on the ecology of lowland river–floodplain systems in the Murray–Darling Basin. Taking a landscape perspective reveals the vast spatial and temporal variability in flooding. We use iconic taxa of the Murray–Darling Basin—river red gum, waterbirds, and fish—to illustrate three strategies for exploiting variable flooding for breeding: (i) reproduce whenever floods occur, regardless of flood characteristics; (ii) reproduce only during floods that provide suitable conditions; and (iii) ignore floods and associated resources because these are too unreliable. Current policy for environmental watering is caught between maintaining individual, ecologically significant sites and a more integrated approach to the management of the riverine system *in toto*. Landscape ecology has a central role to play in informing the management strategy for the Murray–Darling system.

ADVANCES IN ECOLOGICAL RESEARCH VOL. 39
0065-2504/06 $35.00
DOI: 10.1016/S0065-2504(06)39005-8

I. INTRODUCTION

One of the key ideas underlying landscape ecology is the explicit recognition of flows of material and of biota (Forman, 1995). The interactions among different ecosystems mediated by these flows are of crucial importance. With this in mind, a landscape-ecological view of the influence of flooding could hardly be more relevant. While landscape ecology has been criticized as being a "grab-bag of existing ecological and geographical concepts" (Hobbs, 1994; Wu and Hobbs, 2002), landscape ecology also has been endorsed by prominent ecologists as an effective, integrated framework for understanding river–floodplain systems, including flood-associated processes (Ward *et al.*, 2002; Wiens, 2002).

Here, we explore the development of ideas for understanding the ecological processes associated with flooding in the Murray–Darling Basin at landscape scales, review strategies employed by the biota to exploit floods and how this relates to ecological theory, and briefly consider how aquatic production from inundated landscapes may subsidize food webs in surrounding land. We consider how river regulation alters ecological processes associated with flooding. This chapter focuses on the overbank floods characteristic of large, lowland river–floodplain systems because these systems most clearly illustrate the landscape-scale implications of flood regimes (see chapter by Sam Lake, Nick Bond, and Paul Reich for consideration of the ecological role of flooding in headwaters).

II. UNDERSTANDING DYNAMIC SYSTEMS: THE INTEGRATION OF ECOLOGICAL THEORY AND THE LANDSCAPE APPROACH

Riparian areas alternate between being aquatic and terrestrial environments as flow conditions change. This alternation between aquatic and terrestrial phases is more pronounced where large, lowland rivers are fringed by broad floodplains. The fluctuation between aquatic and terrestrial phases creates much heterogeneity in floodplain habitats (Mac Nally *et al.*, 2002). The ecology of the floodplain system cannot be fully understood by considering either the terrestrial phase or the aquatic phase separately from the other (Junk, 1997). Ecologists have struggled to deal with the dynamism of floodplains, especially those of lowlands, perhaps reflecting a human tendency to draw artificial boundaries around systems and research disciplines.

The long history of river regulation in developed countries often is cited as having prevented the study of floodplain ecological processes and propagated a perception of rivers as discrete, two-dimensional conduits running through the landscape (Bayley, 1995; Ward and Tockner, 2001). This view is epitomized by the river continuum concept (RCC) (Vannote *et al.*, 1980), which emphasized the ecological importance of longitudinal, but not lateral, connectivity in rivers. The RCC has limited applicability to large lowland river–floodplain systems, where rivers often flow in multiple, reticulated channels (Brown *et al.*, 1997). One approach to integration of the aquatic- and terrestrial-phase ecology has been to treat the floodplain as a transitional zone (ecotone) between aquatic and terrestrial systems, rather than as an identifiable ecosystem in its own right (Naiman and Décamps, 1990; Gregory *et al.*, 1991; Tockner and Stanford, 2002). Inundation is considered to be a large-scale "external" process that maintains the floodplain ecotone (Naiman and Décamps, 1990). By ignoring internal heterogeneity, this model fails to consider the landscape structure of the river–floodplain system (*sensu* Wiens, 2002).

To develop a theoretical basis for understanding flooding in large river–floodplain systems that recognizes internal heterogeneity, ecologists focused on unregulated tropical rivers. Junk *et al.* (1989) used the extensive floodplain of the unregulated Amazon River as an exemplar to develop the flood pulse concept (FPC)—currently the most influential theoretical framework for floodplain ecology (cf. Ward, 1989). The FPC is based on the principle that rivers and their adjacent floodplains are integrated components of a single dynamic system, linked by strong interactions between hydrological and ecological processes (Tockner *et al.*, 2000). Pulsed flooding is the major factor influencing the biota of river–floodplain systems, with overbank flows determining the degree of lateral connectivity between the river channel and the floodplain, maintaining productivity and increasing biodiversity (see chapter by Sam Lake, Nick Bond, and Paul Reich).

The fauna in the tropical floodplain forests, for which the FPC was conceived, has adapted to a predictable and ancient flooding regime by evolving sophisticated behavioral and physiological traits to deal with inundation (Adis and Junk, 2002). In regulated, temperate systems, flood patterns are less predictable, favoring more opportunistic characteristics in floodplain species (Adis and Junk, 2002). Tockner *et al.* (2000) used a landscape approach to quantify heterogeneity in the floodplains of temperate rivers and, based on their findings, expanded the FPC model to incorporate the less predictable flooding in temperate river–floodplain systems. Flood patterns are most variable in arid and semiarid regions due, in part, to the irregular fluctuations of the El Niño–Southern Oscillation (Walker *et al.*, 1995). Flood patterns in Australian riverine systems are among the most variable in the world. The adequacy of the FPC as a descriptor of the ecology of the river–floodplain systems in the Murray–Darling Basin continues to be explored and debated.

Wiens (2002) contended that adopting a landscape-scale perspective is useful for drawing attention to the dynamic nature of river–floodplain systems, particularly "to the importance of altering the seasonally pulsed connectivity between aquatic and terrestrial ecosystems." By explicitly recognizing human activity, landscape ecology allows ecologists to apply the FPC in thinking about the natural condition of the river–floodplain systems and the continuing impacts of river regulation, as well as to forecasting the outcomes from management that seeks to recreate more "natural" flood regimes.

III. THE LANDSCAPE APPROACH TO UNDERSTANDING THE ECOLOGY OF FLOODS IN THE MURRAY–DARLING BASIN

A landscape approach has been applied to various aspects of the hydrology of the Murray–Darling Basin, including interception of runoff by farm dams (Schreider *et al.*, 2002) and changes in groundwater recharge rates associated with removal of vegetation (Pierce *et al.*, 1993). Given how extensively hydrologists and geologists have studied river–floodplain systems at large scales, there are few examples of use of a landscape approach to understand the effects of flooding on biota in the Murray–Darling Basin. Landscape ecology has been used to understand how flooding and flood-associated processes (e.g., sedimentation) maintain vegetation communities in the Murray–Darling Basin (Bren, 1992; Slavich *et al.*, 1999) and elsewhere (Tiegs *et al.*, 2005). We are not aware of studies in the Murray–Darling Basin of the effect of biota on hydraulics at the landscape scale, although the structure of river red gum forests is likely to influence local flood processes. Fallen timber may influence flow, sediment deposition, and accumulation of nutrients and seeds on floodplains (Boyd *et al.*, 2005). Thick growth of water plants may impede recession of floodwaters from the floodplain wetlands (Barmah-Millewa Forum, 2001).

Comparing different patches within a landscape reveals the importance of geomorphology in determining hydrology and, in turn, ecological processes in river–floodplain systems. Rivers generally are characterized by a longitudinal gradient in channel and floodplain morphology, from steep and narrow in the headwaters to broad and shallow in the lower reaches (see chapter by Sam Lake, Nick Bond, and Paul Reich). Therefore, floodwaters are constrained to the river channel in the headwaters but in the lower reaches overbank flows inundate broad areas of floodplain, creating opportunities for lateral exchange of materials between the river channel and the inundated floodplain.

The geomorphology of each river–floodplain system is a product of its unique and complex geological and hydrological history. For example, the main channel of the Murray River traverses four geomorphic tracts from its headwaters to its mouth, each with a different history (Thoms *et al.*, 2000a; a feature also reported by Currey and Dole, 1978). The geological history of the river tract is reflected in channel capacity and features (e.g., the height of natural levees), which in turn influence flood dynamics. Generally, in river–floodplain systems with natural flooding regimes, the river channel continues to avulse, adding to ecological dynamism and creating challenges for managers with a static mindset (Richards *et al.*, 2002). However, channel migration on the modern Murray River is slow (Gippel and Blackham, 2002).

The volume, timing, and variability of floods may influence ecological processes. The interaction of geomorphology and climate differs substantially throughout the Murray–Darling Basin. When unregulated, the Murray River flooded from snowmelt in early spring while the Darling River was fed by the tropical monsoon in summer. Therefore, the geomorphology and climate of the landscape create the physical template of floods.

An understanding of spatial and temporal variability in connectivity is needed to comprehend river–floodplain systems. A landscape-scale approach is suitable for the study and management of river–floodplain systems because of the high connectivity of these systems, both internally and with the catchment. Water transports nutrients, sediments, organisms (and their propagules) downstream, and also facilitates upstream movement of individuals. Connectivity varies, being minimal in times of drought, when watercourses may be reduced to chains of pools (Bunn *et al.*, 2003). In flood, connectivity is maximized when overbank flows reconnect floodplain wetlands (and floodplains proper) with the main channel (Ward *et al.*, 1999). Homogeneity of the physicochemical properties of floodplain wetlands is maximized when the water level peaks and declines as connectivity decreases (Ward and Tockner, 2001). In the early stages of flooding, waters convey colonists as floodplain wetlands fill, while sustained *in situ* production becomes more important after several weeks of inundation (Jenkins and Boulton, 2003). Floods increase the productivity of river–floodplain systems. Flooding not only controls the growth rate of river red gum but also affects associated populations of folivorous insects (Campbell, 1962; Bacon *et al.*, 1993) and the relative impact of herbivory (Stone and Bacon, 1994).

The effects of flooding on biodiversity are scale-dependent. At smaller spatial scales (e.g., m^2), the challenge created by the habitat alternating between terrestrial and aquatic phases often results in the diversity of taxa being relatively low. At larger scales (e.g., ha), floodplain forests have a mosaic structure and are characterized by high habitat heterogeneity and biodiversity (i.e., β-diversity) (Harper *et al.*, 1997; Tockner *et al.*, 2000). Ballinger *et al.* (2005) demonstrated that flooding promoted diversity of beetles and spiders in Murray River floodplain forests by providing conditions that create a "pulse" in populations of hygrophilic specialists in the short-term, and by creating subtle, ongoing differences in forest-floor conditions, such as the volume of leaf litter. On the Darling River floodplain, Sheldon *et al.* (2002) showed that floodplain wetlands are mosaics, each wetland having a unique biota determined by sporadic connectivity during irregular high flows and other aspects of the hydrological regime.

Flooding has "flow-on" effects for other taxa that may not be regarded as part of river–floodplain systems per se. Inundated floodplains in the

Murray–Darling Basin attract insectivorous birds (Chesterfield *et al.*, 1984; Kingsford, 2000; Parkinson *et al.*, 2002). The density of insectivorous birds is strongly correlated with aquatic-insect emergence, both spatially and temporally (Gray, 1993). Adult aquatic insects have been recorded in the diet of >60 species of Australian birds (Barker and Vestjens, 1940), suggesting that emergent aquatic insects may provide a critical dietary subsidy for terrestrial birds in floodplain forests. Similarly, insectivorous bats use floodplain forest for roost sites more than remnant vegetation in surrounding farmland, possibly because the forest may offer higher prey density than surrounding farmland (Lumsden *et al.*, 2002).

Even during periods of moderate flow, the floodplain provides a "tongue" of relatively mesic habitat that enables species to extend their westerly range into relatively arid areas (Woinarski *et al.*, 2000; Tzaros, 2001). Across a gradient of increasing aridity from east to west along the Murray River, dissimilarity in the structure of riparian habitat compared with adjacent, nonriparian habitat increases as the contrast between the two areas increases (Tzaros, 2001). There is a corresponding increase in the fidelity of birds to riparian habitat (Tzaros, 2001). A corollary is that exchanges of energy or nutrients between the riparian zone and the surrounding terrestrial landscape mediated by birds would be expected to decline as the Murray River flows into more xeric regions in the west. Thus, the flux of energy and nutrients between rivers and the surrounding terrestrial landscape may be influenced by the relative productivity of the two systems (Ballinger and Lake, 2006), and a landscape view suggests that the functional interrelationships between the aquatic and terrestrial ecosystems may differ from place-to-place along the length of the Murray River. Presumably, this also may occur in other major rivers that traverse wide clines in climate.

The flow regime for the Murray River is highly variable over a 50-year time frame, making it difficult to discern patterns in ecological responses to flow in this system (Walker *et al.*, 1995). Perhaps as a result, there are comparatively few "landscape scale" studies relevant to understanding the ecology of flooding in the Murray–Darling Basin, particularly for the dryland rivers (Walker *et al.*, 1995). One theme that has emerged strongly is that the major "pattern" in the biota appears to be a capacity to deal with variability, particularly variable flood regimes.

IV. RIVER–FLOODPLAIN BIOTA IN THE MURRAY–DARLING BASIN: COPING WITH A VARIABLE ENVIRONMENT

Walker *et al.* (1995) contended that a composite of the RCC and the FPC provides the best general model for understanding dryland river systems, if the following modifications are made to the FPC:

- Opportunistic and flexible life-history strategies are considered adaptations to unpredictable, less-seasonal flood regimes.
- The flood pulse has variable duration, rates of rise and fall, magnitude and timing, and this variability has ecological consequences.

While flow regimes in Australian rivers are relatively unpredictable compared with much of the world (Puckridge et al., 1998), the Murray–Darling Basin has a spectrum of flood regimes from reasonably predictable, snow-melt floods in the south of the basin (e.g., Ovens River) to highly episodic flows of the dryland rivers (e.g., Darling River).

Here, we explore how the characteristics of floods interact with the life-history traits of three taxa—trees (focusing on the river red gum *Eucalyptus camaldulensis*), waterbirds, and fish—that are distributed throughout the Murray–Darling Basin, with emphasis on breeding traits. We seek to understand how flooding influences populations of river–floodplain taxa at the landscape scale, and how river regulation is likely to interact with life-history traits to change the distribution of these organisms across the landscape.

There are (at least) three major strategies for exploiting flooding for breeding, summarized in Table 1:

1. Reproduce whenever floods occur, regardless of flood characteristics. This is a high-risk strategy when the characteristics of floods are highly variable, potentially resulting in a low percentage of recruitment (e.g., river red gum).
2. Reproduce only during floods that provide suitable conditions. This reduces the availability of reproductive opportunities, but potentially results in a higher percentage of recruitment per reproductive event (e.g., many waterbirds).
3. Ignore floods and the resources they provide because these are too unreliable. Therefore, reproduce in response to other cues (e.g., some fish in the southern Murray–Darling Basin).

A. Strategy One: River Red Gum *E. camaldulensis*

The river red gum is the iconic tree species of Australian riparian areas. It is the most widely distributed eucalypt in Australia (Boland et al., 1984). For much of its range, particularly the arid central regions of Australia, the river red gum has a serpentine distribution, tracing the margins of temporary and permanent waterways (Boland et al., 1984). The river red gum forms mono-specific, open forests on the extensive floodplains of the Murray–Darling Basin, very different from the diverse stands of Europe and the Americas (Brinson, 1990).

Aborigines may have set fires in the floodplain forests as often as once every 5 years (Lyons, 1988), which may have affected nitrogen dynamics.

Table 1 Characteristics that influence biotic capacity to exploit flooding

Characteristic of flood	River red gum recruitment	Waterbird recruitment (Barmah Forest is taken as example breeding habitat)	Fish recruitment (Ovens River, southern Murray–Darling Basin) (King et al., 2003)
Timing	Winter floodwaters too cool for germination. Prolific germination after spring and early summer floods	Typically breed when food is available (Kingsford and Norman, 2002). In 2000, 8 species bred in spring and 14 in summer in Barmah Forest (Barmah-Millewa Forum, 2001)	Floods most often occur in late winter and early spring (from snowmelt). The majority of the spawning period for 8 of the 12 native fish species in the Ovens River does not coincide with peak flood times
Predictability of flood pulse	Seeds remain viable for years and seedfall is prolific and prolonged	Waterbirds are highly mobile and move to wherever conditions are suitable (Roshier et al., 2001a; Kingsford and Norman, 2002)	Fairly predictable: flooding occurs in spring in 50% of years
Rate of rise and fall	Recession needs to be relatively fast because germinants cannot withstand immersion for prolonged periods	Parents abandon nestlings in response to rapid draw down in water levels (Kingsford and Norman, 2002). Timely, slowly receding waters make ideal feeding grounds (Barmah-Millewa Forum, 2001)	Floods may recede too rapidly. Slow recession necessary to prevent adults and juveniles becoming stranded on the floodplain

Duration of inundation	Germination requires early recession of flooding (i.e., before summer) because germinants cannot withstand total immersion for long periods (Dexter, 1978)	Numbers of nests of colonial waterbirds appear related to areas of river red gum flooded for $\geq$4 months (Briggs et al., 1997)	Estimated time from spawning to full juvenile development for some native fish is 2–4 months, but 60% of floods occur for <10 consecutive days
Area inundated	The best germination usually follows a large spring flood (Boomsma, 1950)	Greater area inundated increases the diversity of habitats available (e.g., for diving ducks, waders)	Greater area inundated increases the diversity of habitats available for spawning and rearing of young
Frequency of flooding	Viability of seeds of 15 years under controlled conditions (Boomsma, 1950). Recruitment at a local scale highly episodic (George et al., 2005)	Recruitment at a local scale (e.g., Barmah Forest) highly episodic	Most fish are long lived compared to flood interval (Mallen-Cooper and Stuart, 2003). In the Ovens River, 6 out of 14 native fish have life spans >3 years
Conclusion	River red gum relies on flooding for germination, but following seasonal conditions (temperature, rainfall) are also influential in the survival of germinants	Waterbirds rely greatly on temporary floodplain wetlands for breeding, but their high mobility enables them to move to where conditions are suitable	Data inconclusive; potentially fish are unlikely to rely heavily on the inundated floodplain for recruitment (King et al., 2003)

Characteristics from Walker et al. (1995), King et al. (2003).

The laminar bark of the river red gum renders it vulnerable to fire. The frequent burning is considered to have maintained a woodland structure (Chesterfield, 1986), described by the squatter Edward Curr as "open, grassy, forest land" with "a very pleasant aspect of mixed Australian and semi-tropical character" [Curr, 1968 (Facsimile of the 1883 publication)]. However, Fahey (1986) quoted reports from 1869 and 1870 stating that the river red gums in Barmah Forest were "so dense that the eye can penetrate only a little way into the forest" and that there were "80–100 trees per acre." Stand density probably varied throughout the forests before extensive intrusion by Europeans. Use of sediment pollen records has not provided a clear picture of densities either (Kenyon and Rutherfurd, 1999; Kenyon, 2001). Forestry records demonstrate that recruitment was episodic even before river regulation commenced (Dexter, 1978).

The floodplain forests have been much reduced in extent since European settlement. In Victoria, there were at least 10^6 ha of grassy riverine forests prior to settlement, but the area now is about 180,000 ha (Department of Natural Resources and Environment, 1997), representing a loss of at least 80% of the original cover.

1. Recruitment Strategy

The river red gum produces many small seeds (Boomsma, 1950). The period of seed fall is extended but peaks in spring in the central region of the Murray River valley, possibly as an adaptation to flooding (Dexter, 1978). Seeds remain viable for up to 15 years (Boomsma, 1950), but ant predation is high (Meeson et al., 2002) and so, turnover of seed banks may be rapid. Seeds may germinate in the absence of flooding in response to extended rainfall. However, without flooding, up to 94% of seedlings may be lost to soil drought within 12 months of germination (Dexter, 1978). Germination is prolific following flood recession in spring and early summer (Dexter, 1978). Seedlings cannot survive immersion for long, so relatively rapid recession of floodwaters is required once germination has occurred. Once floodwaters have ebbed, seedlings may be subject to desiccation in the dry summer conditions or suffer competition from germinated or resprouted grasses (Bren, 1992). Deep floodwaters the following winter or spring may drown yearling seedlings.

2. Opportunism After Recruitment

The river red gum is highly opportunistic in its water use and transpires until water is no longer available. Access to river water, either directly as floodwater or as groundwater, sustains a higher growth rate in river red gum forests than rainfall alone could support (Dexter, 1978). The tree varies its growth rate and canopy foliage in response to water availability. Flooding in spring or summer enhances growth rates (Kayambazinthu, 1997), although (unnatural) prolonged summer flooding may cause tree death.

3. Landscape-Scale Implications

The river red gum is highly opportunistic in its use of water, a limiting and varying resource. This strategy enables the river red gum to be widely distributed, albeit often in a linear fashion along water courses. River regulation has reduced the extent and frequency of winter and spring floods and increased summer floods due to irrigation flows. River red gum forests are in decline due to moisture stress of adult trees and reduced recruitment opportunities associated with a reduction in the extent of flooding, as well as salinity effects. In 2004–2005, over 75% of trees surveyed were classified as stressed (Murray-Darling Basin Commission, 2005). Prolonged summer flooding also has been recognized as a cause of tree death for many years (Boomsma, 1950).

Recruitment in river red gum is highly episodic, reflecting the infrequency of suitable flooding and seasonal conditions (George et al., 2005) and recruitment tends to be patchy, indicating local variation in soil moisture as a result of flood patterns. In some areas, the river red gum is invading floodplain grasslands, apparently in response to altered flood regimes (Bren, 1992). The hydrological conditions in which the river red gum exists in floodplain forests are variable, but conditions for recruitment are relatively specific. Because of the long life span of the species, the basin-wide impact of river regulation on opportunities for recruitment largely has not been adequately considered.

River red gum leaves appear to be a major source of carbon to the Darling River, where flooding regimes play an important role in controlling the input of this carbon to aquatic food webs (Francis and Sheldon, 2002). This differs from arid-zone systems, where production of floodplain carbon appears less important (Bunn et al., 2003). The implications of declining river red gum forests to nutrient cycling and aquatic food webs in the Murray–Darling Basin is an important topic that needs to be explored at the landscape scale.

B. Strategy Two: Waterbirds

Before European settlement, the central Murray area is thought to have supported high densities of aboriginal people relative to less productive regions (Webb, 1984; Lyons, 1988). However, we know of no evidence that hunting pressure played a role in the population dynamics of waterbirds.

Although it is difficult to gauge the populations of waterbirds because of their mobility and aggregation, some reports suggest birds were locally very abundant before river regulation in the Murray–Darling Basin. For example, O'Donohue (1915) wrote of a visit to Lake Hattah: "At least 3000 Black Duck, Teal, and Widgeon were assembled in company with innumerable Black Swans, Pelicans, Ibises, Herons, Maned Geese, Grebes, Cormorants &c. When this concourse of wildfowl was induced to take wing the roar of their pinions and kaleidoscopic movement may be better imagined than described."

Given that populations of waterbirds may be highly variable at any one location, and may be distributed across much of the continent (Roshier et al., 2001a), it is difficult to identify a change in population numbers, let alone attribute change to a specific cause. There is some evidence that populations are in decline (Kingsford et al., 1999). Nevertheless, waterbirds are probably the best understood river–floodplain taxa from a landscape perspective. Annual aerial surveys of waterbirds across much of eastern Australia commenced in 1983 with the objective of determining whether hunting was affecting populations (Braithwaite et al., 1986). These data showed no evidence for an impact of hunting (Kingsford et al., 1999), and constitute the best available large-scale data for assessing effects of river regulation.

1. Recruitment Strategies

Australian waterbirds have climate-driven plasticity in their movements, use of habitat, feeding, and breeding (Kingsford and Norman, 2002; see also chapter by Aldo Poiani). Australian waterbirds vary their reproductive patterns across the Murray–Darling Basin in response to local conditions. Typically, seasonal breeding occurs in areas with regular rainfall and opportunistic breeding in areas with episodic rainfall and flooding. Waterbirds may move thousands of kilometers to find suitable conditions (Kingsford and Norman, 2002 and studies therein). Waterbirds do not necessarily breed in response to heavy rain or floodwaters alone. Instead, they respond to proximate cues, particularly food availability, which may enable birds to assess whether breeding is likely to be successful. Waterbirds abandon nests and nestlings if floodwaters recede too rapidly. Availability of river red gum habitat inundated for ≥ 4 months is a good indicator of the success of waterbird breeding (Briggs et al., 1997), although other, poorly understood factors play a role. When successful breeding is followed by a period of widespread drought, some species may suffer mass adult mortality due to insufficient habitat to support the newly augmented population (Frith, 1962).

2. Landscape-Scale Implications

The plasticity of habitat requirements and high mobility have enabled water-birds to be highly successful, despite the unpredictability of resources and the relative aridity of Australia (Roshier *et al.*, 2001a). The flexibility of life-history traits of Australian waterbirds may insure against sudden population crashes as a result of changes in flow regime at the river-system scale. Rare, very large floods may maintain populations of long-lived species for decades (Kingsford, 2000). We know little about how the capacity of biota to respond to large floods is affected by reduced flood frequency, either as a result of river regulation (Boulton and Lloyd, 1992; Kingsford, 2000) or of climatic change (Roshier *et al.*, 2001b).

The effect of river regulation on waterbird breeding has been modeled for Barmah-Millewa Forest where river regulation was shown to have reduced the frequency of occurrence of conditions for successful breeding by 80% relative to unregulated times (Leslie, 2001). Similarly, breeding opportunities provided by various environmental flow scenarios have been modeled for the Macquarie Marshes (Kingsford and Auld, 2005). Given that dynamics of waterbird populations appear to be regulated at a continental scale, the impact of river regulation on the frequency of breeding opportunities needs to be modeled Australia-wide to quantify the cumulative effect of river regulation on the reproductive ecology of waterbirds.

A sound understanding of the spatial and temporal distribution of critical habitat resources for waterbirds is vital for appropriate management. For example, Victoria and South Australia canceled the 2003 duck-hunting season in recognition of the limited habitat and poor breeding conditions due to low rainfall across eastern Australia. Counterintuitively, waterfowl numbers were expected to be high on wetlands in southern Australia, pro-viding "excellent" hunting conditions. A landscape-scale perspective enabled managers to recognize the high local populations as a concentration of waterfowl in limited refugia, and not a boom in the total population num-bers.

C. Strategy Three: Fish

Early European explorers recorded plentiful, large fish in the rivers of the Murray–Darling system at the time of European settlement (Scott, 2005). The Murray River Fishing Company supplied approximately 150 tonne of (native) fish per year to the Melbourne markets from the 1860s until the 1890s, before the populations of freshwater fish declined to where the fishery no longer was viable (King, 2005 and references therein).

1. Recruitment Strategy

Fish rely on inundated floodplains for breeding resources in other parts of the world (Welcomme, 1979). The applicability of flood-based recruitment models to fish in the Murray–Darling Basin is much debated (Humphries *et al.*, 1999; Graham and Harris, 2005). Fish in the Murray–Darling Basin differ in their life-history traits, such as type of egg and incubation time, so that the capacity to exploit flood conditions for breeding likely differs between species (Humphries *et al.*, 1999). Recent work in the southern Murray–Darling Basin suggests that some native fish may cue on low flows, rather than on floods, because low flows create suitable conditions for breeding, and spawning occurs in main-channel habitats rather than on the floodplain (King *et al.*, 2003; King, 2004; but see Mallen-Cooper and Stuart, 2003). Fish may not use inundated floodplains for breeding in the southern Murray–Darling Basin because the water is relatively cold during most floods, which are fed by snowmelt, and the typical duration of floods (<10 days) is too short (King *et al.*, 2003) (Table 1). Limited use of floodplain habitats also may reflect changed suitability of the floodplain for recruitment due to human activities, such as clearing of riparian vegetation and grazing of livestock (Mallen-Cooper and Stuart, 2003). The golden perch (*Macquaria ambigua*) appears to migrate to spawn in response to rising water levels (Reynolds, 1983; O'Connor *et al.*, 2005), but golden perch recruitment is poor in flood years (Mallen-Cooper and Stuart, 2003). This might indicate a persistence of migratory behavior from a time when the river–floodplain system was better connected hydrologically (Mallen-Cooper and Stuart, 2003).

2. Landscape-Scale Implications

Flood patterns are heterogeneous both geographically, being generally longer and warmer in the north of the Murray–Darling Basin, and temporally, as the volume of rainfall and runoff varies between years. Humphries *et al.* (1999) predicted that this variability may result in different reproductive strategies in fish between different parts of the basin, as well as promoting highly flexible reproductive traits in general. However, whereas waterbirds have been shown to vary the timing of breeding to correspond with favorable local conditions, we are not aware of equivalent data for fish. Similarly, while comparison with "equivalent" species elsewhere has established that Australian waterbirds have plastic reproductive characteristics, to our knowledge such comparisons have not been made between the fish of the Murray–Darling Basin and the fish in regions with highly predictable flood regimes. Basic data on the environmental requirements for fish breeding and recruitment are needed before such questions can be answered. The lack of appreciation of the potential diversity of environmental conditions required by fish for successful breeding may result in simplistic management of environmental flows (Humphries *et al.*, 1999).

V. SYNTHESIS: FLOOD-INDUCED CONNECTIVITY IS CRUCIAL

The elements of river systems are intimately and fundamentally interconnected. The river channel provides longitudinal connectivity. Floods intermittently create lateral connectivity between the river channel and the floodplain. The surface water is connected to groundwater. Last, the movements of animals between riverine and upslope habitats are conduits for exchange of materials between riparian zones and the surrounding terrestrial landscape. Importantly, the extent of connectivity varies through time. In the Murray–Darling river system, the erratic flow conditions often result in the connectivity between the elements being unpredictable.

The biota of the Murray–Darling river system displays a variety of recruitment strategies that may be adaptive in the face of a variable, unpredictable flood regime: prolific germination and high mortality, high mobility to move among river–floodplain systems, breeding in reliable in-channel habitat. For river red gum and waterbirds, we have some understanding of how these strategies interact with hydrological regimes at different scales to determine the distribution of these taxa in the landscape. For native fish, we still lack a systematic understanding of recruitment processes.

In June 2004, the Intergovernmental Agreement on Addressing Water Over-allocation and Achieving Environmental Objectives in the Murray–Darling Basin was signed by the Commonwealth and relevant State Governments. The parties committed funding of Australian $500 million over 5 years to recover 500 GL per annum of water to achieve "specific environmental objectives and outcomes for six significant ecological assets" (Council of Australian Governments, 2004). Five of the significant ecological assets are localities within the river system (e.g., Barmah-Millewa Forest, Hattah Lakes), while the sixth significant ecological asset, the River Murray Channel, forms "an artery between forest, floodplain, wetland and estuarine assets" (Murray-Darling Basin Ministerial Council, 2004). Elevation of the River Murray Channel to the status of a significant ecological asset recognized the importance of longitudinal connectivity for riverine systems.

The Living Murray Initiative augments the environmental water allocations to create seasonal flooding in the large floodplain forests deemed to be significant ecological assets (i.e., Barmah-Millewa Forest, Gunbower, Koondrook-Perricoota Forests). While local agencies are experienced at managing flows in individual floodplain forests, managing for environmental outcomes at the scale of the entire River Murray Channel is new ground and a much more demanding task. The River Murray scientific panel on environmental flows contended that consideration of the impact of management actions on the river and its floodplain at an ever-increasing scale is "essential," but we have insufficient knowledge to do this at a whole-river scale (Thoms *et al.*, 2000b). Thus, bestowing the status of significant ecological asset on the River Murray Channel simultaneously recognizes the need for integrated management of the system and underscores the inadequacy of ecological knowledge to support such an ambitious management regime. It is necessary for ecologists and hydrologists to cooperate to provide information and models to inform landscape-scale management of flows in the Murray–Darling system.

Landscape ecology in the Murray–Darling Basin is still concerned with describing how taxa respond to flooding regimes. Few studies have taken the next step to integrate the responses of different taxa to flooding. We might expect the response of many river–floodplain species to be broadly similar, enabling the pulse in primary productivity created by floodwaters to propagate through the food web as successively higher trophic levels exploit an increased availability of prey. For example, the response of piscivorous water-birds is expected to depend on the response of fish, but the work of King *et al.* (2003) suggested that these taxa might respond to flooding in dissimilar ways, at least in the southern Murray–Darling Basin. A greater understanding of the interconnection of responses to flooding is fundamental to being able to effectively manage river–floodplain systems for ecological sustainability.

ACKNOWLEDGMENTS

We are especially appreciative of the efforts of Aldo Poiani in instigating this volume and are thankful to Carmel Pollino and an anonymous referee for their comments on an earlier version of the chapter. Much of the work of the authors has been funded by the CRC for Freshwater Ecology, Murray–Darling Basin Commission, Australian Research Council, and Hermon Slade Foundation, for which we are most grateful. This is publication number 100 from the *Australian Centre for Biodiversity: Analysis, Policy and Management* at Monash University.

REFERENCES

Adis, J. and Junk, W.J. (2002) Terrestrial invertebrates inhabiting lowland river floodplains of central Amazonia and central Europe: A review. *Freshw. Biol.* **47**, 711–731.

Bacon, P.E., Stone, C., Binns, D.L., Edwards, D.E. and Leslie, D.J. (1993) "Inception Report on Development of Watering Strategies to Maintain the Millewa Group of River Red Gum (*Eucalyptus camaldulensis*) Forests." Technical paper no. 56. Research Division, Forestry Commission of New South Wales, Sydney.

Ballinger, A. and Lake, P.S. (2006) Energy and nutrient fluxes from rivers and streams into terrestrial food webs. *Mar. Freshw. Res.* **57**, 15–28.

Ballinger, A., Mac Nally, R. and Lake, P.S. (2005) Immediate and longer-term effects of managed flooding on floodplain invertebrate assemblages in southeastern Australia: Generation and maintenance of a mosaic landscape. *Freshw. Biol.* **50**, 1190–1205.

Barker, R.D. and Vestjens, W.J.M. (1940) *The Food of Australian Birds: Passerines*, Vol. II. CSIRO, Melbourne, Australia.

Barmah-Millewa Forum (2001) "Report on Barmah-Millewa Forest Flood of Spring 2000 and the Second Use of the Barmah-Millewa Forest Environmental Water Allocation, Spring Summer 2000/2001." Murray-Darling Basin Commission, Canberra.

Bayley, P.B. (1995) Understanding large river-floodplain ecosystems. *Bioscience* **45**, 153–158.

Boland, D.J., Brooker, M.I.H., Chippendale, G.M., Hall, N., Hyland, B.P.M., Johnston, R.D., Kleinig, D.A. and Turner, J.D. (1984) *Forest Trees of Australia*. Nelson Wadsworth and CSIRO, Melbourne.

Boomsma, C.D. (1950) The red gum (*E. camaldulensis* (Dehn.)) association of Australia. *Aust. For.* **14**, 97–110.

Boulton, A.J. and Lloyd, L.N. (1992) Flooding frequency and invertebrate emergence from dry floodplain sediments of the river Murray, Australia. *Regul. Rivers Res. Manag.* **7**, 137–151.

Boyd, L., Mac Nally, R. and Read, J.R. (2005) Does fallen timber on floodplains influence distributions of nutrients, plants and seeds? *Plant Ecol.* **177**, 165–176.

Braithwaite, L.W., Maher, M., Briggs, S.V. and Parker, B.S. (1986) An aerial survey of three game species of waterfowl (Family: Anatidae) populations in eastern Australia. *Aust. Wildl. Res.* **13**, 213–223.

Bren, L.J. (1992) Tree invasion of an intermittent wetland in relation to changes in the flooding frequency of the river Murray, Australia. *Aust. J. Ecol.* **17**, 395–408.

Briggs, S.V., Thornton, S.A. and Lawler, W.G. (1997) Relationship between hydrological control of river red gum wetlands and waterbird breeding. *Emu* **97**, 31–42.

Brinson, M.M. (1990) Riverine forests. In: *Forested Wetlands* (Ed. by A.E. Lugo, S. Brown and M.M. Brinson), pp. 87–141. Elsevier Science Publishers, Amsterdam.

Brown, A.G., Harper, D. and Peterken, G.F. (1997) European floodplain forests: Structure, functioning and management. *Glob. Ecol. Biogeogr.* **6**, 169–178.

Bunn, S.E., Davies, P.M. and Winning, M. (2003) Sources of organic carbon supporting the food web of an arid zone floodplain river. *Freshw. Biol.* **48**, 619–635.

Campbell, K.G. (1962) The biology of *Roselia lugens* (Walk), the gum leaf skeletoniser moth, with particular reference to the *Eucalyptus camaldulensis* (Dehn.) (river red gum) forests of the Murray Valley region. *Proc. Linn. Soc. N.S.W.* **87**, 316–318.

Chesterfield, E.A. (1986) Changes in the vegetation of the river red gum forest at Barmah, Victoria. *Aust. For.* **49**, 4–15.

Chesterfield, E.A., Loyn, R.H. and MacFarlane, M.A. (1984) *Flora and Fauna of Barmah State Forest and Their Management*. Forestry Commission Victoria, Melbourne.

Council of Australian Governments (COAG) (2004) "Intergovernmental Agreement on Addressing Water Overallocation and Achieving Environmental Objectives in the Murray-Darling Basin." Department of the Prime Minister and Cabinet, Canberra.

Curr, E.M. (1968 [Facsimile of the 1883 publication]). The moira. In: *Recollections of Squatting in Victoria, Then Called the Port Phillip District, from 1841 to 1851.* George Robertson, Melbourne, pp. 165–190.

Currey, D.T. and Dole, D.J. (1978) River Murray flood flow patterns and geomorphic tracts. *Proc. R. Soc. Victoria* **90**, 67–77.

Department of Natural Resources and Environment (DNRE) (1997) *Victoria's Biodiversity: Directions in Management.* Department of Natural Resources and Environment, Melbourne.

Dexter, B.D. (1978) Silviculture of the river red gum forests of the central Murray flood plain. *Proc. R. Soc. Victoria* **90**, 175–191.

Fahey, C. (1986) *Barmah Forest: A History.* Department of Conservation, Forests and Lands, Melbourne.

Forman, R.T.T. (1995) Some general principles of landscape and regional ecology. *Landsc. Ecol.* **10**, 133–142.

Francis, C. and Sheldon, F. (2002) River red gum (*Eucalyptus camaldulensis* Denh.) organic matter as a carbon source in the lower Darling river, Australia. *Hydrobiologia* **481**, 113–124.

Frith, H.J. (1962) Movements of the grey teal *Anas gibberifrons* Muller (Anatidae). *Wildl. Res.* **7**, 50–70.

George, A.K., Walker, K.F. and Lewis, M.M. (2005) Population status of eucalyptus trees on the river Murray floodplain, South Australia. *River Res. Appl.* **21**, 271–282.

Gippel, C.J. and Blackham, D. (2002) "Review of Environmental Impacts of Flow Regulation and Other Water Resource Developments in the River Murray and Lower Darling system." Report by Fluvial Systems Pty Ltd to Murray-Darling Basin Commission, Canberra.

Graham, R. and Harris, J.H. (2005) *Floodplain Inundation and Fish Dynamics in the Murray-Darling Basin: Current Concepts and Future Directions—A Scoping Study.* Cooperative Research Centre for Freshwater Ecology, Canberra.

Gray, L.J. (1993) Responses of insectivorous birds to emerging aquatic insects in riparian habitat of a tallgrass prairie stream. *Am. Midl. Nat.* **129**, 288–300.

Gregory, S.V., Swanson, F.J., McKee, W.A. and Cummins, K.W. (1991) An ecosystem perspective of riparian zones. *Bioscience* **41**, 540–551.

Harper, D., Mekotova, J., Hulme, S., White, J. and Hall, J. (1997) Habitat heterogeneity and aquatic invertebrate diversity in floodplain forests. *Glob. Ecol. Biogeogr.* **6**, 275–285.

Hobbs, R. (1994) Landscape ecology and conservation biology: Moving from description to application. *Pac. Conserv. Biol.* **1**, 170–176.

Humphries, P., King, A.J. and Koehn, J.D. (1999) Fish, flows and flood plains: Links between freshwater fishes and their environment in the Murray-Darling river system, Australia. *Environ. Biol. Fish* **56**, 129–151.

Jenkins, K.M. and Boulton, A.J. (2003) Connectivity in a dryland river: Short-term aquatic microinvertebrate recruitment following floodplain inundation. *Ecology* **84**, 2708–2723.

Junk, W.J. (1997) General aspects of floodplain ecology with special reference to Amazonian floodplains. In: *The Central Amazon Floodplain: Ecology of a Pulsing System* (Ed. by W.J. Junk), pp. 3–20. Springer, Berlin.

Junk, W.J., Bayley, P.B. and Sparks, R.E. (1989) The flood pulse concept in river-floodplain systems. *Spec. Publ. Can. J. Fish. Aquat. Sci.* **106**, 110–127.

Kayambazinthu, D. (1997) Effects of flooding regimes on soil moisture dynamics, water balance and growth for a Murray river red gum forest. Ph.D. Thesis,University of Melbourne, Melbourne.

Kenyon, C. and Rutherfurd, I.D. (1999) Preliminary evidence for pollen as an indicator of recent floodplain accumulation rates and vegetation changes: The Barmah-Millewa forest, SE Australia. *Environ. Manag.* **24**, 359–367.

Kenyon, C.E. (2001) Palaeoecological evidence for historical floodplain responses to river regulation: Barmah Forest, south-eastern Australia. In: *Proceedings of the Third Australian Stream Management Conference* (Ed. by I. Rutherfurd, F. Sheldon, G. Brierley and C. Kenyon), Vol. 2, pp. 355–360. CRC for Catchment Hydrology, Brisbane.

King, A.J. (2004) Ontogenetic patterns of habitat use by fishes within the main channel of an Australian floodplain river. *J. Fish Biol.* **65**, 1582–1603.

King, A.J. (2005) Fish and the Barmah-Millewa Forest: History, status and management challenges. *Proc. R. Soc. Victoria* **117**, 117–125.

King, A.J., Humphries, P. and Lake, P.S. (2003) Fish recruitment on floodplains: The roles of patterns of flooding and life history characteristics. *Can. J. Fish. Aquat. Sci.* **60**, 773–786.

Kingsford, R.T. (2000) Ecological impacts of dams, water diversions and river management on floodplain wetlands in Australia. *Aust. Ecol.* **25**, 109–127.

Kingsford, R.T. and Auld, K.M. (2005) Waterbird breeding and environmental flow management in the Macquarie Marshes, arid Australia. *River Res. Appl.* **21**, 187–200.

Kingsford, R.T. and Norman, F.I. (2002) Australian waterbirds products of the continent's ecology. *Emu* **102**, 47–69.

Kingsford, R.T., Wong, P.S., Braithwaite, L.W. and Maher, M.T. (1999) Waterbird abundance in eastern Australia, 1983–92. *Wildl. Res.* **26**, 351–366.

Leslie, D.J. (2001) Effect of river management on colonially-nesting waterbirds in the Barmah-Millewa forest, south-eastern Australia. *Regul. Rivers Res. Manag.* **17**, 21–36.

Lumsden, L.F., Bennett, A.F. and Silins, J.E. (2002) Location of roosts of the lesser long-eared bat *Nyctophilus geoffroi* and Gould's wattled bat *Chalinolobus gouldii* in a fragmented landscape in south-eastern Australia. *Biol. Conserv.* **106**, 237–249.

Lyons, K. (1988) Prehistoric aboriginal relationships with the forests of the riverine plain in south-eastern Australia. In: *Australia's Ever Changing Forests: Proceedings of the First National Conference on Australian Forest History* (Ed. by K.J. Frawley and N.M. Semple), pp. 169–177. Department of Geography and Oceanography, Australian Defence Force Academy, Canberra.

Mac Nally, R., Parkinson, A., Horrocks, G. and Young, M. (2002) Current loads of coarse woody debris on southeastern Australian floodplains: Evaluation of change and implications for restoration. *Rest. Ecol.* **10**, 627–635.

Mallen-Cooper, M. and Stuart, I.G. (2003) Age, growth and non-flood recruitment of two potamodromous fishes in a large semi-arid/temperate river system. *River Res. Appl.* **19**, 697–719.

Meeson, N., Robertson, A.I. and Jansen, A. (2002) The effects of flooding and livestock on post-dispersal seed predation in river red gum habitats. *J. Appl. Ecol.* **39**, 247–263.

Murray-Darling Basin Commission (2005) The year in review 2004–05. In: *Murray-Darling Basin Commission Annual Report 2004–05*, pp. 2–12. Murray-Darling Basin Commission, Canberra.

Murray-Darling Basin Ministerial Council (2004) *The Living Murray Environmental Works and Measures Program*. Murray-Darling Basin Commission, Canberra.

Naiman, R.J. and Décamps, H. (1990) *The Ecology and Management of Aquatic-Terrestrial Ecotones*. Parthenon, UK.

O'Connor, J.P., O'Mahony, D.J. and O'Mahony, J.M. (2005) Movements of *Macquarie Ambigua*, in the Murray River, south-eastern Australia. *J. Fish Biol.* **66**, 392–403.

O'Donohue, J.G. (1915) Wanderings on the Murray flood-plain. *Victorian Nat.* **32**, 26–36.

Parkinson, A., Mac Nally, R. and Quinn, G.P. (2002) Differential macrohabitat use by birds on the unregulated Ovens river floodplain of southeastern Australia. *River Res. Appl.* **18**, 495–506.

Pierce, L.L., Walker, J., Dowling, T.I., McVicar, T.R., Hatton, T.J., Running, S.W. and Coughlan, J.C. (1993) Ecohydrological changes in the Murray-Darling Basin. III. A simulation of regional hydrological changes. *J. Appl. Ecol.* **30**, 283–294.

Puckridge, J.T., Sheldon, F., Walker, K.F. and Boulton, A.J. (1998) Flow variability and the ecology of large rivers. *Mar. Freshw. Res.* **49**, 55–72.

Reynolds, L.F. (1983) Migration patterns of five fish species in the Murray Darling river system, Australia. *Aust. J. Mar. Freshw. Res.* **34**, 857–872.

Richards, K., Brasington, J. and Hughes, F. (2002) Geomorphic dynamics of floodplains: Ecological implications and a potential modelling strategy. *Freshw. Biol.* **47**, 559–579.

Roshier, D.A., Robertson, A.I., Kingsford, R.T. and Green, D.G. (2001a) Continental-scale interactions with temporary resources may explain the paradox of large populations of desert waterbirds in Australia. *Landsc. Ecol.* **16**, 547–556.

Roshier, D.A., Whetton, P.H., Allan, R.J. and Robertson, A.I. (2001b) Distribution and persistence of temporary wetland habitats in arid Australia in relation to climate. *Aust. Ecol.* **26**, 371–384.

Schreider, S.Y., Jakeman, A.J., Letcher, R.A., Nathan, R.J., Neal, B.P. and Beavis, S.G. (2002) Detecting changes in streamflow in response to changes in non-climatic catchment conditions: Farm dam development in the Murray-Darling basin, Australia. *J. Hydrol.* **262**, 84–98.

Scott, A. (2005) *Historical Evidence of Native Fish in the Murray-Darling Basin at the Time of European Settlement—from the Diaries of the First Explorers*. Cooperative Research Centre for Freshwater Ecology, Canberra.

Sheldon, F., Boulton, A.J. and Puckridge, J.T. (2002) Conservation of variable connectivity: Aquatic invertebrate assemblages of channel and floodplain habitats of a central Australian arid-zone river, Cooper Creek. *Biol. Conserv.* **103**, 13–31.

Slavich, P.G., Walker, G.R., Jolly, I.D., Hatton, T.J. and Dawes, W.R. (1999) Dynamics of *Eucalyptus largiflorens* and water use in response to modified water table and flooding regimes on a saline floodplain. *Agric. Water Manag.* **39**, 245–264.

Stone, C. and Bacon, P.E. (1994) Insect herbivory in a river red gum (*Eucalyptus camaldulensis* Dehnh.) forest in southern New South Wales. *J. Aust. Ent. Soc.* **33**, 51–56.

Thoms, M., Suter, P., Roberts, J., Koehn, J., Jones, G., Hillman, T. and Close, A. (2000a) Scientific overview of the river Murray and the lower Darling river. In: *Report of the River Murray Scientific Panel on Environmental Flows: River Murray-*

Dartmouth to Wellington and the Lower Darling River. Murray-Darling Basin Commission, Canberra, pp. 23–54.

Thoms, M., Suter, P., Roberts, J., Koehn, J., Jones, G., Hillman, T. and Close, A. (2000b) Reflections on the process. In: *Report of the River Murray Scientific Panel on Environmental Flows: River Murray-Dartmouth to Wellington and the Lower Darling River.* Murray-Darling Basin Commission, Canberra, pp. 75–79.

Tiegs, S.D., O'Leary, J.F., Pohl, M.M. and Munill, C.L. (2005) Flood disturbance and riparian species diversity on the Colorado river delta. *Biodivers. Conserv.* **14**, 1175–1194.

Tockner, K. and Stanford, J.A. (2002) Riverine flood plains: Present state and future trends. *Environ. Conserv.* **29**, 308–330.

Tockner, K., Malard, F. and Ward, J.V. (2000) An extension of the flood pulse concept. *Hydrol. Proc.* **14**, 2861–2883.

Tzaros, C.L. (2001) Importance of riparian vegetation to terrestrial avifauna along the Murray river, south-east Australia. Masters Thesis, Deakin University, Melbourne.

Vannote, R.L., Minshall, G.W., Cummins, K.W., Sedell, J.R. and Cushing, C.E. (1980) The river continuum concept. *Can. J. Fish. Aquat. Sci.* **37**, 130–137.

Walker, K.F., Sheldon, F. and Puckridge, J.T. (1995) A perspective on dryland river ecosystems. *Regul. Rivers: Res. Manag.* **11**, 85–104.

Ward, J.V. (1989) Riverine-wetland interactions. In: *Freshwater Wetlands and Wildlife* (Ed. by R.R. Sharitz and J.W. Gibbons), pp. 385–400. USDOE Office of Scientific and Technical Information, Tennessee.

Ward, J.V. and Tockner, K. (2001) Biodiversity: Towards a unifying theme for river ecology. *Freshw. Biol.* **46**, 807–819.

Ward, J.V., Malard, F. and Tockner, K. (2002) Landscape ecology: A framework for integrating pattern and process in river corridors. *Landsc. Ecol.* **17**, 35–45.

Ward, J.W., Tockner, K. and Schiemer, F. (1999) Biodiversity of floodplain river ecosystems: Ecotones and connectivity. *Regul. Rivers: Res. Manag.* **15**, 125–139.

Webb, S. (1984) Intensification, population and social change in south-eastern Australia: The skeletal evidence. *Aboriginal Hist.* **8**, 154–174.

Welcomme, R.L. (1979) *Fisheries Ecology of Floodplain Rivers.* Longman Group Ltd., London.

Wiens, J.A. (2002) Riverine landscapes: Taking landscape ecology into the water. *Freshw. Biol.* **47**, 501–515.

Woinarski, J.C.Z., Brock, C., Armstrong, M., Hempel, C., Cheal, D. and Brennan, K. (2000) Bird distribution in riparian vegetation in the extensive natural landscape of Australia's tropical savanna: A broad-scale survey and analysis of a distributional data base. *J. Biogeogr.* **27**, 843–868.

Wu, J. and Hobbs, R. (2002) Key issues and priorities in landscape ecology: An idiosyncratic synthesis. *Landsc. Ecol.* **17**, 355–365.

Effect of Flooding on the Occurrence of Infectious Disease

COLIN R. WILKS, ANDREW J. TURNER AND JOSEPH AZUOLAS

SUMMARY

Changes in climate may affect the incidence of infectious diseases in human and nonhuman animals because they decrease shelter or food supplies forcing populations into closer proximity and because they may decrease host resistance to pathogens. Environmental phenomena, such as floods, may force changes in grazing or foraging behavior that bring populations of susceptible animals into contact with infectious agents that they may have otherwise avoided. Wet and humid conditions associated with floods also provide opportunities for insect vectors of infectious diseases to increase in number and transmit infection more frequently. These underlying principles of infectious disease epidemiology are illustrated with some viral and bacterial diseases that occur in the Murray River region of southeastern Australia.

I. INTRODUCTION

Infectious diseases occur in individuals or in populations of human and nonhuman animals when three sets of factors intersect to produce

ADVANCES IN ECOLOGICAL RESEARCH VOL. 39   0065-2504/06 $35.00
        DOI: 10.1016/S0065-2504(06)39006-X

suitable conditions. These factors are the host, the infectious agent, and the environment.

With respect to the host, it is essential that for disease to occur in an individual, that individual must be susceptible to infection with the infectious agent when the two come into contact. In the case of disease occurring in a population of individuals, as in an epidemic, it is necessary for there to be sufficient susceptible individuals in the population for the epidemic to manifest itself and for the infection to spread from one infected individual to other susceptible individuals. By "susceptible" it is meant that the normal defense mechanisms of the host, such as the immune system and physical barriers like skin or the mucosal surfaces of the respiratory and gastrointestinal tracts, are insufficient to resist the invasion and replication of the agent before it causes damage of varying degree to the infected host. Susceptibility is rarely if ever an all-or-none phenomenon. An individual or a population may be more susceptible when other stressors are prevalent such as intercurrent disease, poor nutrition, inadequate shelter, and overcrowding. Susceptibility is also dependent on the species of the host. Whereas some infectious agents are capable of infecting a wide range of host animal species, some have a very restricted host range, sometimes a single host species.

For infectious disease to occur, it is of course essential that a pathogenic strain of the infectious agent, such as a virus or bacterium, is present. There must be contact between the susceptible individual and a sufficient dose of the infectious agent for the agent to invade, replicate, and cause damage leading to the clinical signs and symptoms that are associated with the particular infectious disease. The damage caused in the host may be the result of destruction of the host's cells in which the agent replicates, as in many viral infections, or the result of toxins and other chemicals produced by the agent, which disrupt the normal functioning of the host's organ systems. Pathogenic strains of infectious agents are able to cause this damage by replicating and releasing their toxins or other factors more rapidly than the host's immune system can counteract them. Also they may have attributes that allow them to resist the effects of the immune system, including the ability to damage critical host cells necessary for specific immune function.

The third factor, the environment, exerts its effects on the other two factors, the host and the agent. Just as one might expect, changes in environmental factors are often associated with changes in the incidence of particular diseases. Extreme climatic conditions that destroy shelter, force people to live closely together in conditions that facilitate the transmission of infectious agents, that destroy food supplies leading to malnutrition, or that provide nutritional and other requirements for infectious agents to increase in number all may lead to an increase in the incidence of infectious disease and indeed to epidemics. Also, since many infectious disease agents rely on biological vectors such as mosquitoes for their transmission, environmental conditions that favor an increase in the population of the vector, typically wetter and warmer seasons, may then lead to an increase in the incidence of the disease in the target host population. Environmental factors may also force changes in geographical location or behavior of the host so that it is placed in closer proximity to an infectious agent and thus more likely to be infected.

In this chapter, we describe the effect of environmental factors, particularly flooding, on the incidence of a number of insect transmitted and other viral diseases, and a bacterial disease, that occur in the Murray River area. Occurrence of the insect transmitted diseases is greater following wetter than usual seasons, whereas the bacterial disease, anthrax of cattle, appears to occur more commonly after prolonged dry periods leading to shortage of suitable grazing so that cattle are forced to graze closer to the ground and forage in dried up waterways, bringing them into closer contact with the spores of the bacterium in the soil.

Even at the time of writing (early January 2006), there are reports of a greater than usual number of people in southern New South Wales and northern Victoria being diagnosed with epidemic polyarthritis, a mosquito-transmitted viral disease described later in this chapter. This follows warm and wet weather conditions in late 2005, which provided ideal breeding conditions for the mosquitoes that transmit the virus.

II. ARBOVIRUS VECTORS AND RAINFALL PATTERNS

Viruses that are transmitted by insects and other arthropods are collectively termed arboviruses. In the Murray Valley region arbovirus epidemics of humans generally occur through the summer and early autumn.

The arboviruses that have caused major disease of humans in the Murray region in the past are *Murray Valley encephalitis virus* (MVEV), a flavivirus, and *Ross River virus* (RRV), an alphavirus. In these instances, the arthropod vector is the mosquito. However, not all species of mosquito are vectors of human arboviruses. Arboviruses have a natural reservoir host/vector cycle in which the reservoir host suffers no apparent ill effects from infection with the virus. Occasionally, the virus breaks out of the natural reservoir host/vector cycle resulting in human infection and even human disease. So human disease is considered to be an "accidental" event and not a part of the normal cycle.

Different species of mosquito have different preferences for hosts on which to feed. For example, *Culex* mosquito species such as *Culex australicus*, a spring breeding mosquito that preferentially feeds on birds and on mammals other than humans, is thought to transmit arboviruses in the bird and animal populations that make up the natural cycle for a particular species of arbovirus. It then diminishes significantly in numbers with the hotter summer weather. However, *Culex annulirostris*, a vicious cross-species feeder and arbovirus vector subsequently becomes the dominant inland mosquito species and brings disease into the human population. The dominance of *Culex annulirostris* is the major reason why arbovirus epidemics occur over the summer period in the Murray region.

Of particular recent concern in relation to human arbovirus disease is the establishment of *Ochlerotatus camptorhynchus* as a major species which in some years becomes the dominant mosquito species around Mildura (34°11′S, 142°09′E) (Azuolas *et al.*, 2003) but is now also being trapped in significant numbers at Swan Hill (35°20′S, 143°33′E). (This mosquito species has been renamed recently as *Aedes camptorhynchus* but the former name is retained in this chapter.) This mosquito is the major vector of RRV responsible for epidemics in Victorian coastal areas. It is usually associated with coastal areas because the mosquito requires pools of salty water for breeding. Its appearance in large numbers in the Murray–Darling Basin (MDB) is thought to have arisen more recently because of the salinity problems along the Murray River, which now provide an ideal habitat for this previously coastal mosquito species. This species has also been shown in the laboratory to be a potential vector of MVEV (Russell, 1993), although there is no evidence to date that it transmits MVEV in the field. This may simply be that until recently the potential vector and the virus did not occupy the same geographical areas. Because it is a cool season breeder, with numbers peaking in spring and wet autumns it could, in conjunction with *Culex annulirostris*, turn what is currently a summer risk period into a spring, summer, and autumn arbovirus risk period. It is also a 24-hour-a-day, cross-species biter compared with *Culex annulirostris*, which is mainly a dusk and dawn biter. Currently, the prophylactic education programs are based on the feeding habits of *Culex annulirostris* so reeducation programs will need to be implemented to prevent the spread of arbovirus disease if, as anticipated, *O. camptorhynchus* becomes a dominant species over a greater range in the MDB.

There is an intimate relationship between mosquitoes and water. Without water mosquitoes cannot breed. The ideal climate for mosquito survival is warm, very humid, and with little, or no, wind. These conditions enable the mosquitoes to survive for relatively long periods of time. This is important for the spread of arboviruses because they are not simply passed mechanically from host to host on the proboscis of a mosquito after it has fed on a viremic host, as in the case of the virus of myxomatosis (see later in this chapter). After the mosquito has taken a blood meal, the arbovirus must replicate and spread through the body of the mosquito and concentrate in the salivary glands before it becomes infective. This process takes about 10 days. In normal years most mosquitoes along the Murray River are not likely to survive long enough to become infective, succumbing to predation or adverse weather conditions. In hot, dry, and windy seasons, there would be very little breeding due to lack of suitable water and rapid desiccation of those mosquitoes that do emerge as adults.

The conditions that favor warm, wet, and humid summers and flooding along the Murray River occur in La Niña years. La Niña occurs when the Southern Oscillation Index (SOI), a measure of air pressure difference between Tahiti and Darwin, is positive. Positive SOI values are associated with stronger Pacific trade winds and warmer sea temperatures to the north as well as cooler waters in the central and eastern tropical Pacific Ocean. Together, these conditions give an increased probability that eastern and northern Australia will be wetter than normal (Bureau of Meteorology, 2005). These conditions are ideal for mosquito breeding and survival, and the proliferation and spread of arboviruses. All sporadic human cases and all epidemics of MVEV and most RRV epidemics have occurred in La Niña years, but there have not been epidemics in all La Niña years.

The popular Forbes (1978) hypothesis maintains that for an epidemic of MVEV to occur in the vicinity of the Murray River, a double season of excess rainfall in all main catchments is required. This means that if rainfall of decile 7 (rainfall not exceeded in 70% of recorded years) or greater occurs in either (or both) of the summer quarters (October–December and January–March) plus the following October–December, then an MVEV epidemic is likely. Forbes concluded that flooding rains in northern and eastern Australia would encourage southerly movement of waterbirds, the natural hosts carrying MVEV, to breed in the Murray region along with heavy breeding of vector mosquitoes.

A. *Murray Valley encephalitis virus* and Kunjin Virus

MVEV and Kunjin (KUN) virus are mosquito-borne flaviviruses endemic to far northern Australia that may cause disease in humans (Marshall, 1988). The illness due to MVEV may be fatal in humans with some survivors experiencing severe ongoing neurological dysfunction. However, the illness caused by KUN virus is not considered to be as severe and is nonfatal. Both viruses sporadically appear in the MDB, usually simultaneously during MVEV epidemics, and independently at other times. KUN virus, but not MVEV, has been detected in sentinel poultry flocks located along the Victorian side of the Murray River in 1990, 1991, 1996, and 2000. Both these viruses may cause encephalitis, but MVEV is considered to be of greater public health concern than KUN virus.

Unfortunately, in an attempt to protect the MDB, the politically correct and confusing term "Australian encephalitis" (AE) was introduced to indicate encephalitis caused by infection with either MVEV and/or KUN virus, which occurs also in other areas of Australia. However, they are different and the diseases are different, so currently the diseases are named after their causative virus (MVEV disease or KUN disease).

MVEV and KUN viruses are arboviruses with a natural cycle involving water birds and marsupials as a vertebrate host and mosquitoes, mainly *Culex annulirostris*, the major vector that brings the virus into the human population. Currently, it is unknown whether there is a hidden focus for these viruses in southeastern Australia or whether they are brought down prior to an epidemic. However, epidemic activity in the southeast has been associated with La Niña events that increase mosquito populations and encourage waterbird migrations and breeding (see chapter by Sam Lake, Nick Bond, and Paul Reich) leading to virus overflowing into human populations. Studies of a wide range of wildlife have shown that the primary hosts of MVEV during years of high virus activity are waterbirds. Pelicaniformes (e.g., cormorants and darters) and Ciconiiformes (e.g., herons) show the strongest evidence of infection (Marshall *et al.*, 1982). Gray kangaroos (*Macropus giganteus*) and red kangaroos (*M. rufus*) also frequently have MVEV antibodies and may play a reservoir host role. Domesticated animals are not thought to be significant as reservoir hosts.

MVEV clinical disease may be a mild febrile illness or include encephalitis and death. Symptoms include sudden onset of fever, headache, neurological dysfunction after a few days, lethargy, drowsiness, confusion, convulsions, fits, coma, and death. Patients that survive encephalitis frequently remain with ongoing neurological disability. Case fatality rates range from 20% to 40% and the apparent to inapparent infection ratio lies between 1:800 and 1:3000.

Epidemics now thought attributable to MVEV infection in humans in Australia were reported in southeastern Australia in 1917, 1918, and 1925 with 114, 67, and 21 cases, respectively, and were named "Australian 'X' disease." Although there is no direct evidence, such as serology, that Australian 'X' disease is in fact MVEV, the presenting symptoms are consistent. The MVEV was isolated from fatal cases during an epidemic of encephalitis in 1951, which was largely confined to the Murray Valley catchments. There were 48 cases (and 19 deaths) during this outbreak. The next epidemic occurred in 1974 giving rise to a total of 58 cases, including 13 fatalities, throughout mainland Australia, with 42 cases recorded mostly in the Murray Valley region. Serological evidence suggests that some nonfatal cases during the epidemic were due to KUN virus infection and that only MVEV was associated with fatal cases.

In those years in the Murray River region when the incidence of clinical disease was low, cases have mostly occurred between mid-February and late March. In those years of severe, widespread epidemics then the cases have occurred from early January through to mid-April.

Occasionally small numbers of infections occur in La Niña years. In 1956, there were two cases of MVEV near Mildura and again, in 1971, a child from Henty near Albury (36°03′S, 146°54′E) became seriously ill with MVEV.

Then in 1974, an MVEV epidemic occurred involving all the mainland states with the majority of cases occurring in the MDB. There have been no recorded human infections with MVEV in the Murray River region since then.

Since 1974, there have been sporadic cases of MVEV reported in northern Australia and in the northwest of Western Australia, especially after heavy rainfall. However, there have not been any cases of MVEV recorded in southeastern Australia. However, in 1984 there was a case of nonfatal KUN encephalitis in northeastern Victoria and two KUN infections were also reported in the Murray Valley. In 1991, there were also two KUN infections reported in southwestern New South Wales.

A direct result of the 1974 epidemic was that arbovirus monitoring programs were initiated in Victoria and New South Wales. In these programs, sentinel poultry flocks are strategically located and bled regularly from November to April to monitor for MVEV activity. The Victorian flocks are placed along the Murray River border with New South Wales. The New South Wales flocks tend to be placed linearly from the Victorian border toward the Queensland border to monitor for southward movement of MVEV coming from Queensland. Mosquitoes are also trapped along the Murray River and screened for arboviruses. It is anticipated that this combination of sentinels and mosquito screening will pick up MVEV activity before it would enter the human population so that there is time to institute preventive measures to avoid human infections.

Three consecutive wet summers with flooding occurred in 1998, 1999, and 2000 with positive SOIs being recorded for those years. During 1998 and 1999 MVEV was regularly detected in northern Australia, and during 2000 it spread south across New South Wales and down to the Victorian border as demonstrated by sentinel poultry seroconversions and domesticated animal serology (Doggett et al., 2001). There were no recorded human cases in New South Wales, quite possibly due to the public warnings given by the Health Department. However, it is quite likely that there were subclinical infections. The following spring of 2000 was wet and warm along the Murray River and there was prolific breeding of mosquitoes; trap catches indicated that Culex annulirostris was the dominant species, making it highly likely that MVEV would continue to spread during the summer of 2001, with the possibility of an epidemic in the human population. However, very high temperatures associated with hot drying winds began in early January, decimating the mosquito populations to a historically low number and commencing the next 4 years of drought. MVEV virtually disappeared except for a seroconversion in sentinel poultry at Menindee (32°21′S, 142°24′E) in 2003, suggesting perhaps a cryptic focus for the virus as opposed to the Forbes hypothesis of it being brought south by waterbirds during times of high rainfall. Or perhaps the Menindee positive serology demonstrated a remnant of the 2000 spread due to a microecological niche that permitted survival of the virus.

Since 1918, if one accepts Australian 'X' disease as MVEV, there have been three major epidemics of MVEV and at least three other summers of known MVEV activity associated with flooding of the Murray River or associated with wet summers and conditions conducive to the survival of mosquitoes, in particular *Culex annulirostris*, the main inland vector of MVEV. Undoubtedly, favorable conditions will occur for the reappearance of MVEV in the future. Active surveillance via sentinel poultry, vector monitoring and virus screening of mosquitoes followed by warnings to the public by health authorities should MVEV be detected will reduce the impact on the human population.

B. *Ross River virus*

Ross River virus (RRV) disease is the most common mosquito-borne viral disease transmitted to humans in Australia with >5000 infections occurring nationally each year (Kay and Aaskov, 1988). It is found in all states of Australia and is most common in the northern states and territories. In the past, the disease has been called Ross River fever, Murray Valley spotted fever, Robinvale disease, and epidemic polyarthralgia, among others. Currently, epidemic polyarthritis is the commonly accepted medical term for the disease caused by infection with RRV. In humans, a wide variety of symptoms and signs may occur from rashes with fever to arthritis that can last from months to years.

The virus has also been isolated from horses with positive RRV serology and exhibiting disease signs such as swollen joints, reluctance to move, ataxia, rash, and poor performance (Azuolas *et al.*, 2003). However, it has so far not been possible to experimentally reproduce this disease in horses by inoculation of RRV. In humans, the disease is characterized by joint pain, typically in hands, feet, and knees, but all joints may be affected. In about 50% of cases a maculopapular rash develops, and fever also occurs in 30–50% of the cases. Severe pain may last up to 3 months with residual arthritis lasting for up to 12 months, and in rare cases for many years. The majority of infections remain symptomless but, extremely rarely, acute severe symptoms resembling meningoencephalitis develop (Russell, 1993).

Transplacental infection occurs in 3–4% of pregnant women; however, there is no evidence to suggest that infections during pregnancy result in abortion or fetal abnormalities (Aaskov *et al.*, 1981).

RRV is found throughout Australia and was first isolated from *Aedes vigilax* mosquitoes trapped near Ross River at Townsville in Queensland.

For most of Australia, peak incidence of disease is through the summer and autumn months, particularly from January through to March, when the mosquito vectors are most abundant in the more tropical regions. However, in southwestern Australia and eastern Victoria, RRV activity may begin in the spring months and peak in early summer then die out because of the reduction in vector numbers due to hostile hot and dry climatic conditions. Outbreaks occur when local rainfall, tides, and temperature provide conditions that increase vector numbers.

In inland regions, the major vector is *Culex annulirostris*, which breeds in freshwater habitats, including irrigated areas. Along coastal regions, salt marsh mosquitoes are the major vectors with *Aedes vigilax* being the northern coastal species and *O. camptorhynchus*, the Victorian coastal species. Serological studies and laboratory investigations have indicated that native mammals, most likely kangaroos and wallabies, are natural hosts for RRV. RRV has been isolated from many mosquito species, indicating wide susceptibility among mosquitoes.

The first reported epidemic of polyarthritis with rash and fever in southeastern Australia occurred in the Riverina region in 1928. Epidemics of polyarthritis have been reported regularly in the Murray River region since the 1950s with the last major epidemic in and around Mildura in the summer of 1997.

It was once thought that RRV was enzootic to northern tropical regions and only occurred sporadically in temperate areas. However, greater awareness of clinical disease by medical practitioners and reliable and highly specific serological testing has shown that epidemic polyarthritis occurs annually in all states. In the MDB, most human cases occur in summer and early autumn paralleling the breeding of *Culex annulirostris*. However, it is common to see a peak of positive RRV IgM serology and clinical disease in horses as early as October and then a second peak when human disease also peaks in summer. The spring peak appears to follow the breeding of *Culex australicus* and gives support to the hypothesis that this species transmits arboviruses in mammals. It also indicates that horses may be excellent sentinels of any forthcoming RRV epidemic in humans.

Serological studies of domestic and native birds have shown very little evidence of infection, therefore birds are not thought to play a role in the transmission cycle of RRV. Mammals are thought to be the most likely reservoir host and serological surveys of domestic and native wild animals strongly favor macropods (wallabies and kangaroos). RRV has been isolated from wallabies (Russell, 1993) and horses (Azuolas *et al.*, 2003).

The epidemic of polyarthritis during the summer of 1997 in the Mildura area was abnormal and is difficult to explain. The summer was very dry and hot, mosquito trap numbers were at a record low, and there were no virus isolations from the trapped mosquitoes. Yet there were approximately 500 reported human cases. However, prior to this epidemic there appeared to be a large increase of clinical disease and positive RRV serology in horses through the winter and spring of 1996 (Azuolas, 1997). This gives further credence to a buildup in a mammalian host prior to entering the human population. But the lack of detectable vectors remains unexplained. It may be that there was an unusual and previously unrecognized vector species of mosquito that was not attracted to the CO_2/light traps used for surveillance of mosquitoes, or there may have been a nonmosquito vector since the apparent winter infection of horses is not consistent with transmission by mosquitoes.

How RRV persists in the cool temperate regions is not understood. There may be a cryptic phase in either vertebrates or invertebrates. Or infected mosquitoes may overwinter to start the cycle again next season. It may be that the virus survives via transovarial transmission; that is, the virus passes from an infected female into her eggs then through the larvae and into the next generation of adults. This may occur in some *Ochlerotatus* species of mosquitoes that lay desiccation resistant eggs. Transovarial transmission has been shown in *O. camptorhynchus* (Dhileepan *et al.*, 1996), which is now a commonly caught species of mosquito around Mildura.

In the Murray Valley region, RRV has been isolated most commonly from *Culex annulirostris*, and also from *Culex australicus*, *O. bancroftianus*, *O. sagax*, *O. theobaldi*, *Anopheles annulipes*, and *Coquillettidia linealis*. The virus has not yet been isolated from *O. camptorhynchus* trapped around Mildura; however, this species is the major coastal vector responsible for epidemic polyarthritis and there is no reason to expect that it will not become a vector in the MDB.

High SOI and La Niña conditions appear to be important for the predictions of epidemics caused by RRV infection in the MDB (Kelly-Hope *et al.*, 2004). When these conditions occur there is a high likelihood of above average rainfall and flooding, or at the very least conditions favorable for the breeding and survival of vector species of mosquitoes in very large numbers. Large numbers of vector mosquitoes do not necessarily mean disease but it does greatly increase the risk.

The social and economic impact of epidemic polyarthritis can be very high in country areas where outdoor work, such as fruit picking, farming, and tourism, are important aspects of the local economy. In these types of industries, workers and customers are exposed to mosquitoes and hence arbo-viruses. Workers cannot work and tourism suffers because people will not come to an area during an epidemic. Thus, the economy of the area suffers.

C. *Japanese encephalitis virus*

A new and emerging threat to Australia is *Japanese encephalitis* (JE). JE virus is a mosquito-borne flavivirus that causes a sometimes fatal encephalitis in humans and horses and may cause abortion and fetal wastage in pigs. It was first isolated in Japan in 1935 from the brain tissue of a person who died with encephalitis. JE virus has spread across Asia, as far as India. It appeared in Australia in 1995, when three residents of Badu Island, in the Torres Strait, developed clinical disease and two subsequently died. Antibodies in humans and pigs were also found on other islands in the Torres Strait, and JE virus was isolated from *Culex annulirostris* collected on Badu Island. This strongly suggests that JE virus may easily become established on the mainland. Pigs were found to seroconvert to JE virus in 1996 and 1997. In March 1998, a boy from the island of Badu was infected with JE virus and later the same month, the first mainland infection occurred in the Mitchell River area on the western side of Cape York Peninsula. Seropositive pigs have also been constantly detected in the Mitchell river area and in the Torres Strait islands since then.

Pigs play a major role as a host species and in the maintenance of JE virus in a region. Therefore, there is real potential for JE virus to rapidly spread down to Victoria via the uncontrolled "feral pig super highway" ranging from far northern Australia down to the MDB and beyond.

III. MYXOMATOSIS

Myxomatosis is a viral disease of rabbits, which causes the development of myxoma-like swellings under the skin and is accompanied by a high mortality rate (Seddon, 1952a).

Although rabbits were introduced into Australia with the First Fleet in 1788 and on numerous occasions thereafter, it was not until an introduction, attributed to Thomas Austin of Barwon Park near Winchelsea in Victoria in 1859, that the pest potential of the species became apparent. The rabbit crossed into New South Wales in 1880 and spread north and west across the continent. Austin killed 20,000 rabbits on his property by 1865 in an attempt to control them. Early attempts at control proved ineffective and even rabbit proof fencing proved impractical to maintain at 100% efficiency (Fenner and Ratcliffe, 1965a).

In 1896, there was an outbreak of a devastating disease in rabbits in South America and further outbreaks occurred in California in North America in 1930. Infective materials derived from rabbits affected with this disease were introduced to Australia in 1926 by Dr. H.R. Seddon, Director of Veterinary Research, New South Wales Department of Agriculture. Laboratory experiments were undertaken that demonstrated the lethality of the virus for Australian rabbits but poor transmission by contact between infected and susceptible rabbits. The matter was left to rest (Seddon, 1952a).

At the request of Dr., later Dame, Jean Macnamara, an expert in poliomyelitis care, to renew the examination of myxomatosis virus for rabbit control, experiments were undertaken at Cambridge, England, by Sir Charles Martin at the Institute of Animal Pathology for Council for Scientific and Industrial Research (CSIR) and Australian quarantine authorities in 1934–1935. It was shown that the virus was highly specific for rabbits and highly lethal by injection but had poor transmissibility by direct contact (Seddon, 1952a; Fenner and Ratcliffe, 1965b).

This work was repeated by personnel of the CSIR Division of Animal Health in Australia led by Dr. Lionel Bull in 1937–1938. It was shown that the virus would not affect domestic or native animals, and field trials were carried out on Wardang Island off the coast of South Australia and later on mainland South Australia. The trials showed that the stickfast flea was efficient at spreading infection between rabbits in an arid environment. It was concluded from the work that myxomatosis might "show some promise of temporary control of a rabbit population where there was an abundance of insect vectors." The work ceased in 1943 at which time it was still not appreciated how important insects would be in spreading myxomatosis (Fenner and Ratcliffe, 1965b).

Further work was taken up with myxomatosis in 1950, again at the insistence of Dame Jean Macnamara, by introducing infection into rabbit warrens at a number of sites in southern New South Wales and northern Victoria. The infection died out at most sites but at a site near Corowa (35°59′S, 146°23′E) the disease initially established then appeared to wane before flaring up again. The reestablishment of the disease was followed shortly by a spectacular epizootic of myxomatosis that in 3 months spread to a large part of New South Wales, southern Queensland, and northern Victoria, particularly along waterways of the Murray–Darling and Cooper's Creek Basins (Seddon, 1952a; Fenner and Ratcliffe, 1965b).

This spectacular spread was in the summer of 1950–1951 when there were floods in southern Queensland and western New South Wales that lead to the development of pest proportions of mosquitoes, sand flies, and other biting insects. The spread coincided with the occurrence of MVEV in humans in the Murray Valley with cases being most numerous in the Mildura and Shepparton areas where myxomatosis was occurring at the same time (Seddon, 1952a). The two converging events led to the obvious public speculation and conclusion that the events were linked with a common cause, myxomatosis virus.

The public concern at the time was such that Sir McFarlane Burnett, Director of the Walter and Eliza Hall Institute, Sir Ian Clunies Ross, Head of CSIR, and Frank Fenner of the Walter and Eliza Hall Institute allowed themselves to be inoculated with myxomatosis virus for cinema newsreel distribution to prove that myxomatosis virus, deliberately released to control rabbits, was not the cause of the encephalitis that was occurring in the same geographical area in humans; it was pre-television and viewing of public events was at picture theatres. As described earlier in this chapter, MVEV was shown to be caused by infection with a different virus although it was also spread by insects.

Myxomatosis ultimately spread to the rest of Australia but the explosive spread in the summer of 1950–1951 was unexpected and provided the basis for the infection to become endemic. The explosive spread was provided for by the floods in southern Queensland and western New South Wales, which in turn provided the conditions for the explosion of blood sucking insects that efficiently transmitted myxomatosis virus from infected to uninfected rabbits. Seasonal conditions for the next 2 years were also favorable for insect populations and the gains in rabbit control from 1950 to 1951 were consolidated in these three seasons. It was fortuitous that the floods occurred in the year when trials restarted (Fenner and Ratcliffe, 1965a).

As a result of the decline in rabbit numbers, sheep numbers boomed, which with the Korean War led to unheard of prices for wool and high incomes for otherwise moderate farmers. Wool prices reached Australian pound for a pound. The wool was required to clothe soldiers at the Korean War during the hostile winter months.

Despite a difficult start, myxomatosis proved to be a cheap and effective means for controlling rabbit numbers and was the major public mechanism for controlling the wild rabbit population until calicivirus was introduced into Australia in the 1990s.

IV. BOVINE HERPES MAMMILLITIS

Bovine herpes mammillitis (BHM) is an ulcerative condition that particularly affects the teats and sometimes the udder of milking cattle. It is caused by infection with a specific bovine herpesvirus. In calves, it causes an ulcerative condition in the mouth. In tropical areas, the same virus causes a generalized infection of the skin known as pseudo-lumpyskin disease.

BHM was first described in Scotland, United Kingdom, and the virus was isolated from the skin lesions. In November 1973, a specimen of a teat skin lesion from a cow in the Cohuna (35°48′S, 144°13′E) area of central northern Victoria was submitted to the Attwood Veterinary Laboratory and a virus was isolated. The virus was shown to be similar to the Scottish isolate establishing its identity as BHM virus (Turner *et al.*, 1974).

Investigations demonstrated that there was infection centered on dairy farms in the Cohuna and Numurkah districts. The affected area was 25 km in radius around Numurkah (36°05′S, 145°26′E) and lesser area around Cohuna reflecting a smaller dairying area. Between 12% and 84% of the cattle in individual herds were affected and the condition was so severe on some cattle that nearly the whole skin of the udder and teats was affected and sloughed. The outbreak occurred over a 6-week period and ceased with the advent of hot and dry weather. By testing blood samples for the presence of specific antibody to the virus, it was demonstrated that yearling heifers had had infection with the virus even though they were not yet part of the milking herd (Turner *et al.*, 1976a,b).

The spring of 1973 was warm and the highest rainfall on record occurred between August and October. The rainfall led to widespread flooding and with mild humid conditions there were very large populations of biting flies and other insects. These insects would have acted as flying inoculation pins that mechanically transmitted virus between infected and susceptible cattle. This would explain why the disease became so widespread and why yearling heifers developed antibody to the infection (Turner *et al.*, 1976b). Tests on yearling heifers in the subsequent year demonstrated that a proportion of them had also become infected with the virus (Turner, unpublished data).

Questioning of farmers in the district revealed that the disease had been seen previously in the early 1950s. When temperature and rainfall records between 1949 and 1973 were checked, there were no similar climatic events to 1973 (Turner *et al.*, 1976b). However, the summer of 1950–1951 was when myxomatosis finally spread and when MVEV occurred in the same general area; it was a time when very large populations of biting insects were in the district.

In February 1979, a generalized skin condition was noted on cattle near Darwin in the Northern Territory, and BHM virus was isolated from skin and blood samples of some animals (St George *et al.*, 1980). February is the tropical rainy season in northern Australia when biting insects are most prevalent. A later study (St George, 1983) showed that many herds, particularly in northern tropical Australia, had cattle with antibody to the virus. Also when sampled in 1977, there were isolated herds in southern Australia with antibody to BHM virus indicating past infection.

It would appear that BHM spreads slowly and probably inapparently in the years between epidemics but disseminates widely in the favorable years when insects and susceptible hosts are plentiful.

V. ANTHRAX

Anthrax is a bacterial disease caused by *Bacillus anthracis* and which gained much public attention when it was associated with the United States anthrax letters in the mail incident in 2001. The disease primarily affects herbivores in which it causes sudden death and blood extravasation from external orifices such as nose, mouth, anus, and vagina. At the time of death, the blood contains very large numbers of bacteria, which transform into spores on exposure to the air. The spores protect the bacterium from the environment providing for its survival in soil between outbreaks of the disease until another animal becomes infected (Seddon, 1952b).

The triggers that lead to anthrax outbreaks occurring are variable and include droughts alone or when heavy rains and floods follow drought and disturb contaminated soil such as at burial sites, old tanneries, or knackeries. The infection is spread by animals having direct or indirect contact with death sites, by biting insects, and through consuming contaminated feed.

Anthrax was introduced to Australia from India probably in bone flour in 1847 at Leppington in New South Wales and again in 1876 at Warrnambool in Victoria. The bone flour was imported as a phosphorus fertilizer for plant crops because Australian soils are deficient in phosphorus. Cattle grazing the land after harvest of the crops contracted anthrax from the spores in the soil (Seddon, 1952b).

The disease spread widely throughout southeastern Australia and became established in an area around the Murray and Murrumbidgee Rivers where the original outbreaks in the 1880s were very severe. Three stations carrying about 220,000 sheep, 4000 cattle, and 600 horses recorded deaths of about 42,000 sheep, 500 cattle, and 60 horses in 1885 and 25,000 sheep, 150 cattle, and 20 horses in 1886. Vaccination was introduced in 1890 using vaccine prepared by people brought out from the Pasteur Institute in Paris. Vaccination greatly reduced stock losses and control was gained over the disease (Devlin, 1943). A locally produced vaccine caused less severe reactions in vaccinated stock and became the only vaccine used in Australia (Seddon, 1952b) until the Sterne strain of *B. anthracis* was used in Australian vaccines from around 1960.

With susceptible stock vaccinated annually, anthrax was absent or caused few deaths, usually in animals that were not vaccinated. Over time, the large outbreaks seen before vaccination did not occur. The disease became associated with small sporadic outbreaks but every so often there would be a large outbreak (Seddon, 1952b). Most outbreaks that occurred between 1937 and 1967 in Victoria were in drought periods or periods with below average rainfall. In the 10 years prior to 1967, there were 47 outbreaks of anthrax in Victoria and all but 2 occurred in the north and northeast. The outbreak occurring in February 1968 involved 27 farms in the Yarrawonga and Shepparton areas of northern Victoria. A total of 85 cattle, and 1 sheep died of anthrax. The reason for the disease occurring on widely separated properties in 1 year is not clear (Flynn, 1969).

Three small sporadic outbreaks occurred in the Shepparton and Yarrawonga areas in the period 1968–1997. In January 1997, a very large outbreak of anthrax started in the Tatura and Stanhope area, which is across the Goulburn River from Shepparton and in an area where anthrax had not been recorded since at least 1914. The outbreak occurred between January 26 and March 26 and 202 cattle and 4 sheep died on 83 properties (Turner *et al.*, 1998a). From February 26, no further properties developed infection and cases stopped occurring on infected farms shortly afterward. There have been a small number of cases developed on infected farms in the 5 years since 1997.

The outbreaks in 1968 and 1997 were controlled by vaccinating all the cattle on infected farms and adjacent areas (Turner *et al.*, 1998b). In 1997, 78,000 cattle were vaccinated on all farms in an area around infected farms and the outbreak was brought under control. The outbreak really developed on February 8–12 suggestive of a common source at a time when daily temperatures and humidity and insect populations were extreme. There had been significant laser leveling of farms in the area to improve irrigation efficiency and this might have disturbed old anthrax grave sites and facilitated contact between stock and the spores as the source of the infection to initiate the outbreak.

All animals that died during the outbreak were burnt and this provided dramatic television images that lead to overseas countries becoming concerned about possible contamination of dairy and meat products coming from the outbreak area. Dairy produce has never been shown to be a source of anthrax and meat produced under Australian conditions has never caused human anthrax. Diplomatic interventions were required to assure trading partners of the safety of Australian export produce.

Since the spore form of the anthrax bacterium can survive in soil for many years, it is likely that outbreaks and sporadic cases of anthrax will occur in grazing animals for years to come when environmental conditions facilitate ingestion of the spores in the soil by grazing animals. In areas where annual vaccination of stock is not deemed warranted, control of these occurrences will depend on early detection of sporadic cases, safe disposal of infected carcasses, and emergency vaccination of at risk stock.

VI. CONCLUSIONS

Floods and droughts are extremes of weather conditions that typify many areas of Australia including the MDB. As well as the many direct effects such events have on human society in the affected area, they also affect the incidence of infectious diseases in animal and human populations, often through the intermediary of insects, whose populations are so intimately dependent on climatic conditions.

Whereas many of these infectious diseases are cause for alarm and the climate is for the most part an uncontrollable variable, there are rational and useful approaches that can be taken. As our understanding of the dynamics of insect populations and their close relationship with certain weather conditions has increased and as we acquire and analyze information from sentinel animals and from direct testing of insects for the presence of particular viruses, we have the ability to predict when outbreaks of disease are most likely to occur. When this predictive ability is coupled with an understanding of the risk factors for infection to occur in individuals, then appropriate actions can be taken to reduce, or even avoid the impact of potential outbreaks of otherwise serious diseases.

REFERENCES

Aaskov, J.G., Nair, K., Lawrence, G.W., Dalglish, D.A. and Tucker, M. (1981) Evidence of transplacental transmission of Ross River virus in humans. *Med. J. Aust.* **2**, 20–23.

Azuolas, J.K. (1997) Arboviral diseases of horses and possums. In: *Arbovirus Research in Australia* (Ed. by B.H. Kay, M.D. Brown and J.G. Brown), pp. 5–7. CSIRO Division of Tropical Animal Science and Queensland Institute of Medical Research, Brisbane.

Azuolas, J.K., Wishart, E., Bibby, S. and Ainsworth, C. (2003) Isolation of Ross River virus from mosquitoes and from horses with signs of musculo-skeletal disease. *Aust. Vet. J.* **81**, 344–347.

Bureau of Meteorology (2005) http://www.bom.gov.au. Accessed December 2005.

Devlin, L.W. (1943) The advent of vaccination of stock against anthrax in Australia: With some of the early history of anthrax in the Riverina district of New South Wales. *Aust. Vet. J.* **19**, 102–111.

Dhileepan, K., Azuolas, J.K. and Gibson, C.A. (1996) Evidence of vertical transmission of Ross River and Sindbis viruses (Togaviridae: Alphaviruses) by mosquitoes (Diptera: Culicidae) in southeastern Australia. *J. Med. Entomol.* **33**, 180–182.

Doggett, S., Clancy, J., Haniotis, J., Russell, R.C., Hueston, L., Marchetti, M., Shaw, C. and Dwyer, D.E. (2001) "2000–2001 Annual Report." Department of Medical Entomology, Westmead Hospital, Westmead, New South Wales.

Fenner, F. and Ratcliffe, F.N. (1965a) The rabbit in Australia. In: *Myxomatosis* (Ed. by F. Fenner and F.N. Ratcliffe), pp. 20–33. Cambridge University Press, Cambridge.

Fenner, F. and Ratcliffe, F.N. (1965b) History and distribution of myxomatosis. In: *Myxomatosis* (Ed. by F. Fenner and F.N. Ratcliffe), pp. 1–9. Cambridge University Press, Cambridge.

Flynn, D.M. (1969) Anthrax in Victoria. *Vic. Vet. Proc.* **27**, 32–33.

Forbes, J.A. (1978) *Murray Valley Encephalitis 1974 also the Epidemic Variance Since 1914 and Predisposing Rainfall Patterns.* Australasian Medical Publishing, Sydney.

Kay, B.H. and Aaskov, J.G. (1988) Ross River virus disease (Epidemic polyarthritis). In: *The Arboviruses: Epidemiology and Ecology* (Ed. by T. Monath), Vol. IV, pp. 93–112. CRC Press, Florida.

Kelly-Hope, L.A., Purdie, D.M. and Kay, B.H. (2004) El Niño Southern Oscillation and Ross River virus outbreaks in Australia. *Vector-Borne Zoo. Dis.* **4**, 210–213.

Marshall, I.D. (1988) Murray valley and Kunjin encephalitis. In: *The Arboviruses: Epidemiology and Ecology* (Ed. by T. Monath), Vol. III, pp. 151–190. CRC Press, Florida.

Marshall, I.D., Woodroofe, G.M. and Hirsch, S. (1982) Viruses recovered from mosquitoes and wildlife serum collected in the Murray Vallet south-eastern Australia, February 1974, during an epidemic of encephalitis. *Aust. J. Exp. Biol. Med. Sci.* **60**, 457–470.

Russell, R.C. (Ed.) (1993) *Mosquitoes and Mosquito-Borne Disease in Southeastern Australia.* Department of Medical Entomology, University of Sydney and Westmead Hospital, New South Wales.

Seddon, H.R. (1952a) Myxomatosis. In: *Diseases of Domestic Animals in Australia* (Ed. by H.R. Seddon), Part 4, pp. 177–183. Commonwealth Government Printer, Canberra.

Seddon, H.R. (1952b) Anthrax. In: *Diseases of Domestic Animals in Australia* (Ed. by H.R. Seddon), Part 5, Vol. 1, pp. 8–40. Commonwealth Government Printer, Canberra.

St George, T.D. (1983) A serological survey for neutralizing antibodies to bovid herpesvirus 2 in cattle in Australia. *Aust. Vet. J.* **60**, 187–189.

St George, T.D., Uren, M.F. and Melville, L.F. (1980) A generalized infection of cattle with bovine herpesvirus 2. *Aust. Vet. J.* **56**, 47–48.

Turner, A.J., Kovesdy, L., Cianter, M.L., Nicholls, W.A. and Chatham, R.O. (1974) Isolation of bovine herpes mammillitis virus from dairy cattle in Victoria. *Aust. Vet. J.* **50**, 578–579.

Turner, A.J., Kovesdy, L. and Morgan, I.R. (1976a) Isolation and characterization of bovine herpes mammillitis virus and its pathogenicity for cattle. *Aust. Vet. J.* **52**, 166–169.

Turner, A.J., Morgan, I.R., Sykes, W.E. and Nicholls, W.A. (1976b) Bovine herpes mammillitis virus of dairy cattle in Victoria. *Aust. Vet. J.* **52**, 170–173.

Turner, A.J., Galvin, J.W., Rubira, R.J. and Miller, G. (1998a) Anthrax explodes in an Australian summer. *J. Appl. Microbiol.* **87**, 196–199.

Turner, A.J., Galvin, J.W., Rubira, R.J., Condron, R.J. and Bradley, T. (1998b) Experiences with vaccination and epidemiological investigations on an anthrax outbreak in 1997b. *J. Appl. Microbiol.* **87**, 196–199.

Some Economics of Floods

JOHN FREEBAIRN

SUMMARY

The chapter evaluates decisions of households, businesses, and governments to locate in flood-prone areas, to invest in flood mitigation projects, and to choose risk management strategies. In general, it will be sensible for some to locate in flood-prone areas, and even after investment in flood mitigation it will be sensible for some costs of flood damage to be incurred. Well-informed households and businesses locating in flood-prone areas will do so only if on balance the costs of flood damage are less than the benefits in the good times. Governments have key roles of providing information and of coordinating decisions on flood mitigation investment projects. However, a government policy of subsidizing those adversely affected by floods is criticized on both efficiency and equity grounds.

I. INTRODUCTION

Floods are one of the natural hazards which reduce national income and economic opportunities in Australia. While most of the fundamental causes of floods lie beyond human control, decisions on where to locate houses and

ADVANCES IN ECOLOGICAL RESEARCH VOL. 39   0065-2504/06 $35.00
DOI: 10.1016/S0065-2504(06)39007-1

businesses, on investment in different forms of flood mitigation, on warning systems, and on risk management strategies can reduce the loss of property, income, and lives. This chapter is concerned primarily with the choice of decisions by individuals, businesses, and governments in the context of probable floods so as to maximize individual household and business economic outcomes and the economic prosperity of the country.[1]

Several sets of estimates are available of the costs of floods to the Australian economy over recent decades. These costs include the loss of property and contents, the loss of production, deaths, and injuries. Over the period 1967 through 1999, the Bureau of Transport Economics (2001) estimated flood-caused reductions in the value of property, contents, and production (in 1999 Australian dollars) of $10.4 billion, or an average of about $300 million a year. Over the same period, the study estimated there were 99 deaths and 1018 serious injuries attributable to floods. Smith (2002) estimates there are about 150,000 households subject to one in a 100-year flood at an average annual cost of about $150 million, and twice as many households are at risk of more extreme floods. Most of the economic costs are associated with severe and infrequent floods.

Economic decisions are affected by the prospect of floods, and as is the case for other economic decisions, decisions about floods aim to maximize household well-being or utility given limitations or constraints imposed by available labor, capital, natural resources, and technology. Because of scarcity relative to needs, choices have to be made. Floods, droughts, and other natural disasters are largely climate-driven probabilistic events which provide a part of the natural resources constraint set facing individuals, businesses, and governments. At least three sets of decisions can be noted. First, the frequency and magnitude of flood damage varies widely from one location to another, so that household and business location decisions need to take account of differences in relative returns, costs, and flood probabilities at alternative locations. Second, the magnitude of flood damage at any particular location can be mitigated by investment in dams, levees, building quality, drainage systems, and other infrastructure. But, these investments have opportunity costs, and decisions need to balance the benefits and costs of different forms of flood mitigation and of their different levels. Third, risk averse households and businesses often explore the choice among different risk management decision options, including insurance, to even out the costs of the unpredictable and often large but infrequent costs of flood damage.

---

[1]The general methodology and results follow related economic analyses of other natural disasters, and in particular droughts. Examples of Australian studies are Freebairn (1983) and Botterill (2005).

The rest of the chapter explores in more detail these different decisions for responding to floods so as to improve economic well-being.

II. FLOODS AND LOCATION DECISIONS

Because the potential damaging effects of floods vary widely by location, the location choices of households and businesses take into consideration the relative probabilities of floods of different sizes and the magnitude of damage caused. For example, households in a city or town adjacent to a river might be located on the floodplain and be subjected to flood damage on a regular basis or they might locate on the foothills with a lower through to a zero probability of flooding. Again, farms producing crops and live-stock might be located on areas of land subject to flooding every few years or on surrounding areas of infrequent, less damaging, or no flooding. In many cases, in addition to differences in the probabilities and damage caused by floods, the attraction of the different location options vary also in terms of other valuable characteristics such as proximity to shops and schools, the ease of and costs of construction and maintenance of buildings and other structures, to differences in agricultural productivity, and so forth. Individual households and businesses in choosing where to locate will take into account the differential incidence and effects of floods, along with other factors, affecting the relative benefits and costs of the alternative locations.

To illustrate how differences in the probabilities and costs of floods, and of other factors bearing on costs and returns, affect location decisions consider a simple two area location choice where the probability of flooding differs between the two areas. The two areas are designated by subscript i, with $i = 1$ for a low-flood-prone area and $i = 2$ for a high-flood-prone area. The probability of a large and a small flood in area i is P_{ij}, where j designates the degree of flood damage with $j = 1$ for a large flood and $j = 2$ for a small flood, and for our example $P_{11} < P_{21}$. The economic benefits and costs of location i under climatic (or flood-prone) condition j is given by $B_{ij} - C_{ij}$. Here the benefits B_{ij} might include rental value in the case of a home and agricultural product sales in the case of a farm, and the costs C_{ij} include the usual production-operating costs plus any damage costs associated with a flood.

Now, in making a location choice, the risk neutral and fully informed household or business will compare the expected net benefits from each of the location choice options, which we designate as $E(L_i)$ namely, for location $i = 1$

$$E(L_1) = P_{11}(B_{11} - C_{11}) + P_{12}(B_{12} - C_{12}) = \Sigma P_{1j}(B_{1j} - C_{1j}) \qquad (1)$$

and for $i = 2$

$$E(L_2) = P_{21}(B_{21} - C_{21}) + P_{22}(B_{22} - C_{22}) = \Sigma P_{2j}(B_{2j} - C_{2j}) \qquad (2)$$

To maximize his or her economic position, the individual decision maker will choose location 1 when $E(L_1) \geq 0$ and $E(L_1) > E(L_2)$, and location 2 if $E(L_2) \geq 0$ and $E(L_2) > E(L_1)$.

The illustrative two location and two flood state model can be generalized to encompass many locations by letting $i = 1, 2, \ldots, n$ locations, and for many flood states by letting $j = 1, 2, \ldots, m$ states of different severity of flooding. Additional information is required for the larger sets of benefits, B_{ij}, costs, C_{ij}, and probabilities, P_{ij}, to compute the expected net benefits, $E(L_i)$.

The private decisions of individual households and businesses to locate or not in high-flood-prone locations, and decisions on the particular activities to locate in these areas, which maximize their individual expected net benefits will in most cases also approximate those choices which maximize society well-being. The correspondence of individual choices with society choices follows if there are no market failures. An important proviso is that the private sector decision makers have access to all available information on the probabilities of flooding, P_{ij}, including the various dimensions of the extent and severity of floods by location. Since such information has public good properties[2] there is a compelling case for government intervention to ensure the provision of such information. Ignoring the options of investment in flood mitigation infrastructure, which we discuss in Section III of this chapter, differences in the probabilities and severity of floods across regions per se do not generate other forms of market failure arguments in the form of external costs[3] and monopolistic market power[4] to warrant government intervention in private sector decisions to locate in flood-prone areas.

------

[2]By public good properties I mean (i) non-rival consumption so that my use of a piece of information does not reduce its availability to other users and (ii) high costs of exclusion so that once the information is produced and made available to one person it is difficult and costly to deny its use by others. With public goods such as defense, law and order, and much information there is a tendency for everyone to free ride on others, and as a result a private market system will produce and consume too little of the public good when assessed against national or social benefits and costs.
[3]By external benefits and costs is meant spillover benefits and costs to third parties who are not involved in a market transaction between buyers and sellers.
[4]By market power I mean that the individual firm faces a downward-sloping demand curves for its product and it can alter the market price by changing the quantity it sells. In these cases, firms will choose to produce and sell less of their product than would maximize social welfare.

A number of important implications for the efficient allocation of society's scarce resources to flood-prone locations can be drawn from the simple illustrative example of this section. First, if land was plentiful and returns and costs other than those associated with floods were identical across different locations, no household or business would locate in a flood-prone location. But, in reality land is limited and we now consider this realistic context. Second, activities whose returns are least affected by the ravages of floods will locate in the more flood-prone locations and those activities whose benefits and costs are very sensitive to floods will avoid flood areas. Third, and associated with the second point, the rental return on land will be inversely related to the degree of expected flood damage. In practice, the different rates of rent return to land in the different locations will tend to equate the expected net returns of Eqs. 1 and 2. Fourth, in general, because of land scarcity, some activities will be located in flood-prone areas and it will be in society's economic interest to incur some flood damage costs because the benefits of location there exceed the costs, inclusive of flood damage costs.

Another implication of the foregoing analysis is that the decision to locate or not in a flood-prone area, and the definition of a flood-prone area to be avoided, including by government regulation, should be based not on a specific flood probability but rather on the basis of an expected net benefit payoff as illustrated in Eqs. 1 and 2. A not uncommon local government regulation is to restrict buildings to areas with less than a one in a 100-year-flood chance. So long as the expected net benefit is positive, society gains by locating activity in the area. Depending on circumstances described by differences in the benefit, cost, and probability terms, differences which will vary across areas at any one time and also these differences can vary over time for any one location, the expected net benefit $E(L_i)$ may be positive, and warrant location in flood-prone areas and some flood damage, for flood probabilities of one in 5, 20, 100, 200, or whatever years.

III. FLOOD MITIGATION STRATEGIES

So far, the analysis has assumed that nature alone determines the frequency and severity of flooding. However, in practice, society can take decisions which effectively reduce the costs imposed by the extreme variability of rainfall. By investing in dams, levee banks, drainage systems, the robustness of buildings, including height above the ground and structural strength, and other infrastructure, the frequency and severity of floods can be modified. The costs to society of floods when they occur can be in part mitigated by early warning systems which enable households and businesses to move people, house contents, and livestock to higher ground and so to reduce the losses and costs caused by floods. An important set of decisions involves assessing from the perspective of the economy, the time path of the economic costs and benefits of different flood mitigation strategies so as to determine those options which improve the overall allocation of society resources, and the society maximizing flood mitigation investment levels by flood area. These strategies involve an up-front investment in infrastructure which reduces the severity of floods in the future and reduces the magnitude of costs caused by flood damage. Such projects improve society economic performance if the discounted value of expected future benefits exceeds the opportunity costs of the investment.

One formal model or method of analysis for evaluating the economic effects of flood mitigation investment strategies is the benefit–cost model. As an illustration, Ng (1992) describes a specific application to evaluate investment in urban drainage to reduce flood damage to buildings and to the loss of life. A benefit–cost model assessment of each flood mitigation investment proposal involves several steps and requires numerous sets of information.

The first step is to describe the project proposal, for example a dam, levee bank, building regulation, or early warning system. Second, the effects of the investment on the probability distribution of flood outcomes, the P_{ij} terms of Section II, and the changes in flood damage, the B_{ij} terms in Section II, have to be determined. This stage requires hydrological and other physical and biological data. For computational purposes, the simplest assumption is a constant percentage reduction in each of the flood state probabilities. However, for the majority of flood mitigation investment strategies it is more likely that the investment will reduce most of the losses associated with the small and medium flood states, but with a smaller reduction in the magnitude of damage associated with the less frequent and more damaging extreme flood events.

Given the physical effects data, the third step involves setting out a table of the time path of the economic costs and benefits of the flood mitigation proposal. Most of the extra costs are those of the investment outlays incurred in the initial period, plus any ongoing repairs and maintenance expenses. Since flood mitigation in many cases changes the natural flow of water, particularly if the investment includes dams or levee banks, for these projects the investment will also alter ecological and environmental outcomes along the water course. Changes to these outcomes need to be measured, initially in physical and biological terms, and then dollar equivalent values need to be placed on the environmental changes. Bennett (2005) provides an overview and critique of related market and of nonmarket contingent valuation and choice modeling methods for placing a dollar equivalent amount on changes in those environmental outcomes for which there is no market. To the extent that the environmental outcomes are adversely affected, they represent an ongoing economic cost of the flood mitigation investment proposal.

Benefits of the flood mitigation proposal are the expected reduction in costs caused by floods, where the costs saved include fewer deaths and injuries, less expenditure on restoring flood-damaged buildings and contents and other structures, and less loss of production. These expected cost savings extend for each year over the life of the flood mitigation investment. Most of the costs are an up-front investment cost, and the benefits are ongoing over the life of the project.

The fourth step of the benefit–cost model assessment of a flood mitigation investment project is to aggregate the time paths of different costs and benefits of step three into a single measure. This involves using a discount rate factor reflecting both time preferences for consumption and the rate of return on investments across the economy to express dollars in the current and future years of the project in comparable present value terms. Then, if the net present value of benefits less costs exceeds zero, the project adds to national economic well-being and should be adopted; and if the net present value is negative, the project should not be accepted. Alternatively, the project should be adopted if the ratio of the present value of benefits to the present value of costs exceeds unity.

Inevitably, there will be uncertainty about most parts of the benefit–cost model assessment. For most projects, there will be imperfect information on the magnitude of flood reduction effects of a flood mitigation investment, the dollar values to place on the cost savings, the extent and value of environmental effects, and the magnitude of the discount rate, and often even the construction cost of the initial investment is uncertain. One strategy is to rerun the benefit–cost analysis for different potential values, or scenario studies, to evaluate the robustness of the estimated net present value or benefit–cost ratio.

For many, the faith by benefit–cost analysts that everything has a monetary value, or can be given a monetary value, is regarded as a heroic assumption at best and more likely foolhardy or worse. Particular areas of concern in evaluating flood mitigation investment projects are the value of life, of changes to the environment, and anxiety associated with the threat of floods. Rather than use a money metric, another set of the literature proposes that a multiattribute evaluation system be used. Prato (2003) provides an example evaluation for comparing and evaluating five project options for management of the Missouri River system. This category of models uses the basic framework and same information of the benefit–cost model, except that monetary values are not attached to the nonmarket effects on lives saved, changes to environmental outcomes, and other nonmarket changes. Expert-based methods, such as the Delphi method, the nominal group technique, or simulation methods (described in, for example, Seaver, 1976 and the references in Prato, 2003) are used to derive weights for the different market and nonmarket attributes, and then to weight the multiple attributes to achieve an index for ranking alternative flood mitigation investment options.

Investments in most flood mitigation strategies will be subject to both declining marginal social benefits and to rising marginal social costs over the relevant decision levels. For example, ever larger dams and levees involve increased costs per unit flood reduction, and the additional flood reduction probability saves less and less flood damage costs. Ideally, the level of any particular mitigation investment will be chosen where marginal social benefits approximate marginal social costs.

One important implication of the marginal cost and benefit properties of flood mitigation investments is that seldom will it be optimal to invest to the point where all flood damage is mitigated.

Most, but not all, investments in flood mitigation will require some intervention by governments. In most cases, investment in infrastructure to reduce the frequency and damage caused by floods to protect one household or business will have external benefits in reducing flood damage for other households and businesses in the flood-prone area. Seldom will it be feasible or low cost for the different actual, let alone potential, beneficiaries to form clubs to internalize the benefits of large flood mitigation infrastructure projects. Further, most of these projects also involve both economies of scale and natural monopoly characteristics. The few exceptions with no or very few instances of significant market failure include large land developers and the structure of individual buildings. Then, in the majority of flood mitigation infrastructure investments, government intervention to internalize the externalities and to reap the society advantages of economies of scale will be necessary to achieve the levels that maximize economic efficiency.

However, the market failure arguments for government intervention to ensure that flood mitigation investments which pass a benefit–cost evaluations are undertaken does not extend to government funding of these investments. This would constitute a subsidy and income transfer to households and businesses locating in flood-prone regions at the expense of households and businesses in the rest of the economy. One effect of a subsidy is that businesses and households in the flood-prone area benefit via reduced costs of flood damage but they do not meet the costs of the investments. The subsidy will encourage a reallocation of economic location from the non-flood-prone areas to the now more profitable flood-prone areas, resulting in a loss of national productivity and real income. One option is for government in the first instance to fund flood mitigation investments, and then to pay for these investments with an additional fee or charge on land in the better protected flood locations. Given that the flood mitigation investment project provides net benefits, the gains to individual households and businesses from reduced flood damage costs exceed the additional land fee, and government breaks even.

IV. RISK MANAGEMENT

So far, it has been argued that in many flood-prone areas, it will be in the interest of national economic efficiency for some households and businesses to locate in these areas, even after investment in various flood mitigation strategies, because the expected benefits exceed the expected costs. These households and businesses face the prospect of unpredictable flood events which, on occurrence, can result in large current period losses. Many but not necessarily all households and businesses understanding the vagaries of nature will establish and implement risk management strategies to cover periods of temporary large losses in the event of flood damage. Risk management strategies could include setting aside contingency funds, borrowing, or taking out flood insurance, and thereby transferring the risk to others. Risk averse decision makers are willing to pay insurance premiums which exceed the expected losses (and insurance payouts) from floods.

The development of flood insurance markets has proved difficult in most countries, and in particular in Australia where at most 10% of potential flood victims are estimated to be covered by flood insurance (Smith and Handmer, 2002). Optional flood insurance schemes include bundling flood, fire, theft, and other losses into the one product (the UK model), treating flood damage as an optional insurance add-on component, or a government underwritten and operated program (with the United States being one example). Flood insurance programs are confronted by the problems of inadequate generally available information on the probability distribution of floods, and especially about the damage caused at the individual household and firm level. The combination of general information uncertainty together with asymmetry of information between particular clients and the insurer results in adverse selection and moral hazard, both of which work to undermine the development of a market for flood insurance as compared with, say, fire insurance.

V. GOVERNMENT POLICY

Governments have a natural inclination to provide financial and other support to those adversely affected by floods and other natural disasters. In part, they respond to dramatic and selective media presentations of the damage caused, and to a community sense of supporting those in need, especially if their source of plight is considered beyond their control. From an economic perspective, government provision of natural disaster assistance is both a cause of economic efficiency loss and it is a relatively blunt instrument for meeting social equity goals.

In earlier sections, it has been argued that well-informed households and businesses who locate in flood-prone areas do so because on average the good times more than compensate for the losses in the event of floods, and most also arrange mechanisms to smooth over the good and the bad times. In these senses, those locating in the risky flood-prone locations are making similar decisions as other risk takers in businesses with uncertain markets and new technology and by individuals in career choices with such risky careers as sports, the arts, and dotcom areas. Further, the utility- and profit-maximizing decisions of individual households and businesses on where to locate was argued also to correspond with the allocation which would maximize the efficient allocation from the perspective of society. Adding the opportunity to invest in flood mitigation strategies provides the same general efficiency result.

Then, what is the nature of the adverse efficiency effects of government assistance to flood-affected households and businesses? In essence, the assistance is a form of subsidy, funded by taxes on the rest of the community, which reduces the downside costs of locating in a flood-prone area while allowing the recipients to keep all the benefits in the good times. Crudely, it is a policy of privatizing the wins and socializing the losses. In terms of our more formal analysis of Eqs. 1 and 2 in Section II, the expected net benefit of locating in the relatively flood-prone location is raised relative to that of the relatively flood free location. This subsidy has two adverse efficiency effects.

First, by making the flood-prone areas relatively more attractive, the subsidy induces some households and businesses to move from low flood risk locations to higher flood risk locations. The distortion of location decisions represents a misallocation of resources and a loss of national productivity in much the same way as selective tariffs for some products but not for others, and other selective subsidies, distort economic choice decisions. Ironically, by encouraging more people and businesses to locate in flood-prone areas the subsidies also lead in the future to additional flood damage (and government handouts).

A second adverse efficiency effect of flood assistance subsidies is that they reduce the incentive for individual households and businesses to develop contingency strategies to cope with the inevitable floods. That is, they are encouraged to spend all the gains during the good times and to rely on government handouts in the event of floods. This effect partly explains the difficulty in establishing a market for flood insurance.

Governments have a strong mandate for redistribution to meet society goals of equity and fairness. In Australia, there is an elaborate system of progressive income taxation and of means-tested social security pensions and allowances to meet equity goals, and key services, such as education and health, are provided either for free or at a subsidized price to all. While not without criticism, the tax and social security system is a relatively direct and effective system for redistributing from the better-off to the more disadvantaged. By contrast, subsidies for those affected by floods are less targeted and effective in meeting the needs of the disadvantaged. In general, flood assistance payments are about the same per household regardless of family income, and payments to flood-affected businesses, like any subsidy paid to businesses, is an indirect and a crude way to provide assistance to disadvantaged families.

Current practices of many local governments to restrict by regulation the location of houses and businesses in flood-prone areas and/or to enforce building codes with reference to minimum floor heights and building standards to withstand floods are debatable. Arguably, all the benefits and costs faced by individual decision makers over these choices also correspond to social benefits and costs, that is, there are no external costs or benefits. Then, fully informed households and businesses will make the society optimum choices with no need for government intervention. Clearly, local (and other) governments should provide all available information on flood probabilities and damage to individual decision makers so that they are fully informed, and at least as informed as government. Individuals then have the comparative advantage over governments of tailoring that information for making decisions with reference to their own specific objectives and situations using private information not available to government officials.

For areas where the combination of information about the probabilities of floods and the damage they cause is not available or is very crude, and for the instances of rare but very expensive flood damage, it might be argued that governments should provide a minimal level of support from general taxation revenue for such rare and exceptional flood damage as "a society insurer of last resort." For this combination of circumstances, private insurance markets are unlikely to develop until better information is collected. As argued above, such catastrophic natural disaster relief does represent a subsidy to those locating in the flood-prone affected areas and results in a misallocation of national resources. However, with such relatively rare events it might be argued that few households or businesses factor the subsidy into their decisions so that in practice the efficiency losses are small. If and where these arguments are accepted, challenges for government are to clearly spell out the nature of extreme events when government provides a last resort safety net, then to stick to those guidelines, and to leave the private sector to incorporate the probabilities of "nonextreme" flood episodes into their decision choices.

There are at least two important and noncontroversial areas for government policy intervention affecting floods. The first area is the collection and dissemination of information on the probability distribution of floods in their different dimensions such as frequency, water velocity, height, duration, and other characteristics which affect flood damage. This is likely to be an ongoing process with the prospects of climate change and as new flood mitigation investments and changes in land use alter water flows. Such information has classic public good characteristics which mean that it would be under-provided if left to competitive markets, and government impartiality can be important.

A second area of market failure and area for government intervention is in the provision of investments in flood mitigation. In most cases externalities and scale economies result in market failures. Government involvement usually would include planning and evaluating infrastructure investment options, and in implementing a system of land charges or fees on beneficiaries to recoup the investment costs. The actual construction and maintenance activities could be tendered out or undertaken by governments themselves.

VI. CONCLUSIONS

Floods are a given characteristic of the economic environment for some locations, and they impose constraints on the decision options available to households, businesses, and governments. For the investment of labor and capital, infrastructure projects in dams, levees, early warning systems, and so forth can be undertaken to modify the environment and to reduce the damage costs of floods, but seldom will complete elimination of floods be justified. For some households and businesses over time, the benefits exceed the costs, including flood damage costs, of locating in flood-prone areas, and it is in the economic efficiency interests of the economy both to locate some activities in flood-prone areas and to bear the damage costs caused by occasional floods.

A mixture of government intervention and the self-motivated decisions of private households and businesses is required in making decisions to locate in flood-prone areas and to invest in flood mitigation. Households and businesses fully informed with government-provided information on the probability distribution of floods will choose to locate or not in flood-prone areas based on a comparison of expected benefits and costs in locating in different areas. Those choosing flood-prone areas, or areas facing the risks of other types of natural disasters, also will choose risk management strategies to enable them to balance the bad times against the good times. Government subsidies, including flood damage grants, distort location choices in favor of too many locating in flood-prone areas, the subsidies reduce the incentives to use the windfall gains of the good times to offset the losses caused by floods, and they represent a relatively blunt instrument for meeting society's concerns for equity and fairness.

Governments have key roles to play in the planning and financing of flood mitigation investment projects. The importance of external benefits and scale economies with the projects are important market failures. Each proposal should be subjected to a formal evaluation, whether that be a benefit–cost assessment or a multiple attribute assessment. Either mode of assessment will require considerable technical and economic input. For efficiency reasons, those who benefit through lower flood damage costs should be asked to meet the investment project costs, with an additional levy or charge on land being a good option.

ACKNOWLEDGMENT

With the usual caveats, I gratefully acknowledge the comments on an earlier version by Harry Clarke and Aldo Poiani.

REFERENCES

Bennett, J. (2005) Australasian environmental economics: Contributions, conflicts and 'cop-outs'. *Aust. J. Agric. Resour. Econ.* **49**, 243–262.

Botterill, L. (2005) *From Disaster Insurance to Risk Management: Australia's National Drought Policy.* Springer, The Netherlands.

Bureau of Transport Economics (2001) "Economic Costs of Natural Disasters in Australia, Report 103." Bureau of Transport Economics, Canberra.

Freebairn, J. (1983) Drought assistance policy. *Aust. J. Agric. Econ.* **27**, 185–199.

Ng, Y.-K. (1992) Optimal investment in urban drainage: A framework for cost-benefit analysis. *Aust. Econ. Rev.* **3**, 19–28.

Prato, T. (2003) Multiple-attribute evaluation of ecosystem management of the Missouri river system. *Ecol. Econ.* **45**, 297–309.

Seaver, D. (1976) "Assessment of Group Preferences and Group Uncertainty for Decision-Making." Research Report, University of Southern California, Social Science Research Institute.

Smith, D. (2002) The what, where and how much of flooding in Australia. In: *Residential Flood Insurance: The Implications for Floodplain Management Policy* (Ed. by D. Smith and J. Handmer), pp. 73–98. Water Research Foundation of Australia, Canberra.

Smith, D. and Handmer, J. (Eds.) (2002) *Residential Flood Insurance: The Implications for Floodplain Management Policy.* Water Research Foundation of Australia, Canberra.

Tiddalik's Travels: The Making and Remaking of an Aboriginal Flood Myth

JOHN MORTON

SUMMARY

Many flood myths have been told throughout the length and breadth of Aboriginal Australia. In the past 100 years or more, these stories have increasingly found non-Aboriginal audiences. This chapter examines one particularly popular flood story, "Tiddalik the Frog," and situates it in both classical and postclassical contexts. I illustrate how the moral import of the story has shifted as a result of changes in presentation and audience, particularly since the story was deemed most suitable for children. However, I argue that the story contains more mature lessons, as relevant today as they were in the past, regarding environmental management and the sustainable use of water.

I. INTRODUCTION

"Flood" seems like such a straightforward description, but perhaps a flood is more complicated than it might at first appear. To understand fully the meaning of floods in Australia is to engage a veritable flood of metaphors and associations in the English language connected with overwhelming excess—what we might call "thinking white." But if we want to know what Indigenous Australians have been thinking about floods during their time in Australia, then we need to engage the philosophies that have come to be known collectively as "The Dreaming" or "The Dreamtime." We need, as

ADVANCES IN ECOLOGICAL RESEARCH VOL. 39   0065-2504/06 $35.00
DOI: 10.1016/S0065-2504(06)39008-3

anthropologist Bill Stanner said many years ago, to begin to "think black" by "not imposing Western categories of understanding" (1979, pp. 24–25).

However, "thinking black" is not a simple proposition. For one thing, Aboriginal people have a very long history in Australia, long enough for there to have been a great many social, environmental, and religious changes and upheavals. Most of what we know today about classical Aboriginal cultures relates to but a brief period before and after official European colonization of the country in 1788. The rest is not history, but archaeology, although part of what we know about the long haul is that Aboriginal people have had to adjust to dramatic changes in climate and geography (Jones, 1994). Indeed, some of the most familiar aspects of the current shape of Australia are the result of recent flooding in the Late Quaternary (Jennings, 1971). By and large, Aboriginal oral traditions relate to the familiar landscape that we now know, notwithstanding the many changes that have occurred in that landscape since the coming of European settlers.

Variation in time is matched by variation in space, for it is estimated that, in 1788, some 250 different Aboriginal languages were spoken in Australia (McConvell and Thieberger, 2005, p. 78), devolving into a great many significantly different dialects. If language is taken as an index of culture, then it is evident that "thinking black" has meant different things at different times and in different places. While it is now customary to align "prehistory" with the era prior to European invasion and settlement of the continent, and think about it as an "eternal Dreamtime" in which Aboriginal people were enmeshed, "primitive," uniformly "traditional," and lacking in "development" (Wolfe, 1991), this view is little more than a colonial conceit.

To get past the conceit, it is necessary to treat Aboriginal flood mythology with all due care and caution. We cannot expect a straightforward answer to questions like: "What do Aborigines think about floods? How do floods figure in Aboriginal mythology?" There is no one flood story, just as there is no single Aboriginal culture and no singular Dreamtime. There are only stories which come from particular places and which have been narrated and recorded at particular times and on particular occasions. The very definition of myth is also complicated by this diversity, for we find that Aboriginal stories contain elements which are both historical and fantastic. Flood stories may preserve the memory of real events, but those events are routinely embellished by human imagination. Moreover, a story might not simply be told as a linear narrative. It may be revealed in layered parts or in alternative versions, some open, others secret. It may not be told at all, but rather sung as poetic verse, dramatized through performance or given symbolic expression through the plastic arts. And if it is told, it may be told for various reasons— for amusement, as warning, as homiletic instruction, or to bring to life spiritual agency and thereby effect transformations in identity, because what you know can define who you are (a man, a woman, a child, a member of one clan rather than another, a member of one "tribe" rather than another).

To do flood stories justice, therefore, it is necessary to say something about what they are for. But because many (though far from all) of the stories were originally recorded without due regard to the significance of Indigenous context, it is often possible to say more about the context of their reception into Australian society at large and less about what Aboriginal people thought about them when they were first transcribed. Indeed, one might argue that, in Australia today, the context of Aboriginal mythology is as much "white" as it is "black." Such complexity makes it difficult to write briefly and systematically about Indigenous flood mythology. The most one can hope to do is to sample the corpus and give some general idea about a range of themes that might be more thoroughly examined.

I approach these matters through an analysis of a single myth: "A Legend of the Great Flood" (Thomas, 1939, pp. 26–30), which is one version of a story more generally known as "Tiddalik the Frog" (or something similar).

II. THE MYTH OF TIDDALIK THE FROG

I summarize the story as follows:[1]

There was once a terrible drought. The land was parched and even the most reliable waterholes had dried up. Many died. The situation was so critical that those who survived resolved to bury all their differences and meet in peaceful council to discover what was causing the problem. Creatures traveled far and wide to get to the meeting—kangaroos, wallabies, bandicoots, koalas, possums,

---

[1] I summarize the story in my own words partly in order to save space. However, it is important to point out that this is methodologically justifiable. Many theorists (Lévi-Strauss, 1970, pp. 12–13; also see Stanner, 1989, pp. 84–85; Berndt and Berndt, 1988, pp. 389–390 specifically in relation to Aboriginal mythology) have drawn attention to the fact that myths characteristically exist in multiple versions which are, as it were, constantly moving, transforming themselves into something else, yet at the same time remaining continuous with earlier versions. The way in which I rewrite the Tiddalik story here, so long as the revision is in some sense faithful to the original, actually mimics the manner in which the story might otherwise be transformed in other contexts. As Lévi-Strauss points out (1968, pp. 216–217), a myth consists of *all* its versions, and none of them is original in anything other than a historical or sequential sense. That is to say, in an analytical sense, no one version can claim priority as the *true* or *real* version, although such claims for particular versions are certainly made on political or ideological grounds. This is partly what is meant when it is said that myths are "timeless." They are grounded in the past, but entirely of the present, which is largely why Malinowski (1926) defined them as pragmatic charters.

emus, brolgas, bell birds, butcher birds, brown snakes, echidnas, and others besides.

Seabirds came from the coast and traveled far inland. They discovered that an enormous frog had swallowed all the water in the land, thus giving rise to the drought. The meeting decided that, if the land was to be rescued, they would need to find a way to make the frog laugh, which would cause all the water to be expelled from his body.

The creatures gave the job to Kookaburra, who rose to a tree and, looking the bloated frog in the eye, began his characteristic laughing call. His laugh grew louder and louder, until the landscape resounded with his call. The other animals watched the frog intently, but he just impassively blinked his eyes.

The job was then given to Frill-Necked Lizard, who expanded his frill, puffed out his jaws, and capered up and down. Still the frog sat impassively.

So Brolga was given a chance. She danced and danced in front of the frog, but still he did not move. The situation was serious.

The animals began to call in a confused cacophony and fight among themselves. Some fought over food and some tried to consume other creatures. Everything became very chaotic.

But then Eel suggested that he could do the job. The other creatures were skeptical, but when Eel began to writhe and contort into different shapes, the frog opened his eyes and burst out laughing. The laugh was like thunder and water teemed from inside him, filling the rivers and waterholes, and covering the land. Only the highest peaks remained dry and many were drowned.

Pelican, whose feathers were then entirely black, sailed in a canoe between the peaks to rescue the survivors. Eventually he came to a heavily populated island. He rescued all the men from the island, but, in spite of her pleas, always refused to evacuate one particular woman, because he wanted her as his wife. Since she did not want to go with the Pelican, she filled her possum skin rug with a log so that it looked like she was still asleep and escaped into the bush.

When Pelican returned, he was furious at being duped and sought vengeance. He painted himself with white clay and set out to kill her male relatives. However, the first pelican that the avenger met was so shocked by the avenger's white appearance that he struck the avenger with a club, killing him outright. Since that time, all pelicans have been both black and white, in memory of the flood.

When the flood subsided, the country was fresh and green once more. And when the morning broke, a new flood emerged—a flood of golden sunlight.

We know very little about how this story might have been told in Aboriginal communities. A search through libraries and the internet reveals that the story is now thought almost exclusively to be a document most suitable for children, although most contemporary versions of the flood story invariably stop short at the point where Tiddalik releases the world's water from his body: nothing is said about Pelican's final interventions; nor is there normally any other extension of the story. Rather, the newer stories conclude at the moment Tiddalik breaks the drought and causes the flood. Such, for example, is the case with the version transcribed and pictured as "Tiddalik the Flood-Maker" in the popular coffee table *Dreamtime Book* (Roberts and Mountford, 1973, pp. 24–25).[2]

How, then, do we identify the general significance of the flood motif as it travels from old to new contexts? One important clue lies in the way in which the stories are *totemic*, in this case mainly using animal characters acting in quasi-human form. Aboriginal religion has been generally characterized as totemic (Stanner, 1979, pp. 106–143), in the sense that it uses "nature" (in this case, animal species) as a symbolic resource for the creation of mythic characters and scenarios. But Aboriginal totems are not exclusively faunal. Some are plant species; a number are either purely human in form, or

---

[2]Contemporary versions of the story include books (or chapters thereof, including scripts for plays), video recordings (including dramatizations), sound recordings (including musical renditions), and e-publications (including poetic verse and school children's illustrated retellings). I have also been privately informed of oral versions yet to be widely disseminated. Of the written versions, the earliest I have found are from the Kurnai (Gunai, Ganai) people of Gippsland (Curr, 1887, pp. 547–548; Howitt, 1888, pp. 54–55; also Howitt, 1904, pp. 486–487; Smyth, 1878, pp. 429–430; Vanderwal, 1994, p. 37), who located the story in features of the landscape in and around Port Albert (Massola, 1965)—a matter I return to in more detail in the conclusion to this chapter. It seems likely that most subsequently recorded versions of the story, as told by both Aboriginal and non-Aboriginal people, have been directly or indirectly influenced by the Gippsland versions, either through new narration (Reed and Hart, 1965, pp. 175–177; Roennfeldt, 1981; Troughton, 1994) or by the influence of a newly heard version on a preexisting one. Ellis (2004, pp. 15–17), for example, recently recorded a Wonnarua version of the story from the central coast of New South Wales— one which appears to be related to a local site near Wollombi (http://www.aboriginal-hunter.com/dreaming.php3?story=1) and to be the basis for the Tiddalick's Toothy Tale Oral Health Project, which uses the local story "to encourage preschool children to drink water between meals and wash their mouths with water after meals to reduce the chance of tooth decay" (http://www.hnehealth.nsw.gov.au/news/releases/2004/July% 2004/hunter_22.htm). One of Tiddalik's more recent and interesting migrations was to Warwick in southeastern Queensland, where in 1998 he appeared as a 9-tonne granite greenbelt mascot sitting in Riverside Park (http://www.flexi.net.au/greenbelt/Pages/tid-dalik.html). Similarly, Tiddalik is now a mascot for the Amphibian Research Center (http://frogs.org.au/arcade/product_info.php?cPath=37&products_id=78&PHPSESSID= 66lacc9bfd3f162f918ad14e52a87e42). It is also clear that Tiddalik is related to different but in some sense allied stories told in Aboriginal communities throughout a large part of Australia (Parker, 1953, pp. 36–37; Henry, 1967, pp. 19–20; Timaepatua and Gumudul, 1977; Unaipon, 2001, pp. 25–41, 53–59).

at least quasi-human; and a few are inanimate objects. Yet others are natural forces or phenomena—stars, the sun, the moon, wind, fire, or water, for example. Naturally, it is the totemic character of water which is of most concern to us here.

In the story outlined above, water is associated with the frog known as Tiddalik, a name which is almost certainly onomatopoeic and based on the frog's call. In other parts of the continent, water is associated with other beings, such as Goanna[3] or Dirrawong ancestor of the Bandjalang people of northeastern New South Wales (Steele, 1983, p. 3), the well-known Rainbow Serpent or Ngalyod of the Kunwinjku people of western Arnhem Land (Berndt and Berndt, 1970, pp. 20–24), or the equally famous "spaceman-like" ancestral Wandjina of the people of the Kimberley region of Western Australia (Mowaljarlai and Malnic, 2001). In yet other areas, totemic water is, as it were, a law unto itself. In central Australia, for example, Arrernte people tend to speak of the relevant Dreaming simply as *Kwatye* (Water, Rain, Clouds), and the rain ancestors as *Atwe Kwatye* (Water People) (Strehlow, 1907, pp. 25–28).

Totemism is an aspect of Aboriginal *cosmology*—part of Aboriginal people's general theory about how the world was made, how it continues to fit together, how it should continue to do so, and how human beings were, are, and will be part of those processes. In the most general of terms, this is what is subsumed by the term "Dreaming" or "Dreamtime." This is how the anthropologist Bill Stanner famously defined this Indigenous concept:

> The Dreaming is many things in one. Among them, a kind of narrative of things that once happened; a kind of charter of things that still happen; and a kind of logos or principle of order transcending everything significant for Aboriginal man. If I am correct in saying so, it is much more complex philosophically than we have so far realised. I greatly hope that artists and men of letters who (it seems increasingly) find inspiration in Aboriginal Australia will use all their gifts of empathy, but avoid banal projection and subjectivism, if they seek to honour the notion (1979, p. 24).

Our appreciation of the meaning of totemism, and of The Dreaming, is now much refined, although Stanner's salutary injunction to avoid "banal projection and subjectivism" still very much applies.

As a cosmology, The Dreaming, in all its shapes and forms, has systematic qualities, generally subsumed in Aboriginal English under the rubric of

---

[3]The Lace Monitor, *Varanus varius* (Cogger, 1979, pp. 241–242).

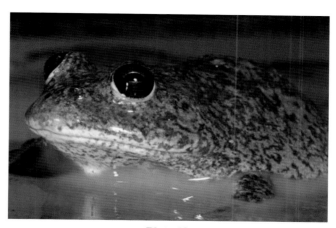

Plate 12

Plate 13

"Law."[4] Rose (1998a) has defined these qualities in the context of the Yarralin (Ngarinman and Ngaliwurru) peoples, who live on the northeastern extremes of desert country in the Northern Territory. She abstracts from Yarralin cosmology a number of axioms and principles.

> ... all parts of the cosmos are alive. ... Each part is related to other parts, but in the beginning each part walked separately. It is on this basis that Dreaming Law is a law of autonomy ... And while each part is autonomous, each is equally part of a total system. The goal of the system as a whole is to reproduce itself as a living system, while the goals of each part are: (1) to reproduce itself as a part; and (2) to maintain the relationship between itself and other parts ...
>
> These relationships are crucial, for without them life cannot continue. ... Moral rules concern these critical relationships between parts of the system. These relationships are based upon principles which I identify as those of response, balance, symmetry and autonomy. Each part is autonomous as a fundamental feature of the integrity of life. Each part balances and is balanced by other parts. In order for parts to be balanced they must be symmetrical, that is, of equal power ... And in order for balance to be maintained, parts must communicate. They do this by acting (being alive) and responding. In actions, parts assert their autonomy and strength; in responding, parts delimit the boundaries of other parts and thus implicitly their own boundaries. Each part of the cosmos is thus seen as a moral agent and in behaving morally each part reproduces the relationships through which the whole system continues to enhance life (1998a, pp. 242–243).

Those (and there are many) who see in this philosophy a blueprint for an appropriate form of ecological consciousness in Australia (and elsewhere) often speak of Aboriginal "wisdom" in relation to "the environment" (Knudtson and Suzuki, 1992). It is often forgotten, however, that it is not a blueprint for an unchanging Eden or perfect harmony. Rather, it is about harmony as a *goal*, and sometimes as an *achievement*. There would be no reason for goals, let alone struggles to achieve harmony, if the cosmos were not in some sense imperfect and prone to become out of kilter. This is why Stanner found at the heart of Aboriginal religion the idea that life is fatally flawed—"a joyous thing with maggots at the centre" (1989, p. 37).

A flood, by definition, occurs when the world gets "out of kilter." A flood is an excess—an excess of water, but it can also be, an excess in a much more general and extended sense. It also implies an absence, because an excess obscures and

---

[4]Sometimes rendered textually as "lore." However, the notion of Law (a body of rules and regulations) as opposed to lore (a body of stories and folk knowledge) is the correct original coinage in Aboriginal English.

obliterates other possibilities. In Thomas's published version of the Tiddalik story, for example, the flood displaces another excess—a drought. But as the story progresses, the opposition between an excess of fire (the sun's heat) and an excess of water (the flood) is resolved to arrive at a median state, a replenished countryside which is once again returned to a green, productive, and pleasingly sunlit land. It is not entirely inappropriate, therefore, that a group of Commonwealth Scientific and Industrial Research Organization (CSIRO) scientists should have recently adopted the name Tiddalik to describe their "irrigation area model for predicting and managing drainage return flows" (Hornbuckle *et al.*, 2005), for it would seem that, in Aboriginal terms at least, Tiddalik has been conceived as just that—a "model" for dealing with the proper and productive control of water in an otherwise parched land. On the other hand, Tiddalik alone is hardly an *ideal* model for an environmental management system given his role in the definition of excess.

But who precisely is this creature? What does it mean to say that he might be, in Rose's terms, an autonomous agent needing to respond to the agency of others in a world where human beings seek to establish symmetry and balance through appropriate response? Lévi-Strauss (1966) coined the phrase "the science of the concrete" to describe how certain kinds of scientific thinking can be contrasted to allegedly typical analytical procedures of Western science. Those who operate in a concrete mode, sometimes called "analogical reasoning," do "not move abstractly and hierarchically from axiom to theorem to corollary," but rather "construct theories by arranging and rearranging, by negotiating and renegotiating with a set of well-known materials" (Turkle and Papert, 1990, p. 136). The observations on which analogical reasoning is based are no less acute than those based on systematic scientific experimentation, but in the situations referred to by Lévi-Strauss empirical procedures tend to be based on knowledge which is pitched at the level of "sensible intuition" (Lévi-Strauss, 1966, p. 15), requiring the development of generally heightened senses of touch, sight, smell, hearing, and taste. Aboriginal people living in classical or neoclassical conditions are well known for such sensual mastery; witness, for example, their famed tracking skills. In these terms, it is important to understand that Tiddalik is not simply a fantasy (although he most certainly is that). He is, as already intimated, an analogue model applied theoretically to the workings of a sector of the world—the world of water.

Frogs are obviously connected to water, but, as far as I can ascertain, nobody has precisely identified the species which is the inspiration for this model, even though some have given a general idea. For example, Mudrooroo writes in connection with the Tiddalik story:

> Frogs are associated with water and in dry arid regions in times of scarcity of water, there is a frog which gorges itself with water then buries itself into the ground and waits, perhaps for years, for the next

rainfall. The [Aboriginal] people know this and when there is a drought, they dig them up for water (1994, p. 67).

There are in fact a number of burrowing frogs in Australia, some of which are literally "water-holding." One is actually called the water-holding frog[5] and is distributed across a large swathe of arid Australia (Fig. 1). On a more general canvas, whenever Australian landscapes are flooded, frogs of all types appear in numbers.

In the Legend of the Great Flood, the breaking of the drought is a way-station along the road to conflict resolution and harmony. That is why the other creatures in the story are so desperate to make Tiddalik laugh. But why laughter? Laughter, of course, normally signals happiness and social harmony, but nobody is happy or harmonious during a drought. Laughter is also a release from anxiety. Resources become scarce in a drought; not just water, but all those plants and animals which depend on water diminish in numbers. Severe lack of water causes worry and concern. It also throws creatures into conflict as they compete for scarce resources and invade each other's territories in search of sustenance. But the release of water is by no means a straightforward release of anxiety, for it prompts yet more.

Figure 1   The water-holding frog *Cyclorana platycephalus.* Lydia Fucsko/frogs.org.au (See Color Plate 12.)

---

[5]*Cyclorana platycephalus* (Cogger, 1979, p. 51). However, neither the water-holding frog nor any of the genus *Cyclorana* is currently found in Gippsland, Victoria (Cogger, 1979, pp. 48–51), the area from which the Tiddalik story (as we now know it) appears to have originated.

The flood destroys and, having already suffered the effects of a parched landscape, yet more people die. If each part of the cosmos is autonomous and ideally balanced, a severe action leading to imbalance requires an equally severe reaction to restore balance. Excess breeds excess, and for Aborigines, as Rose suggests, this is fundamentally a moral proposition.

I noted earlier that the Tiddalik story is now most often told to children. This is usually to teach them the virtues of sharing and the dangers of greed. Tiddalik is not only a master of water, when the chips are down, he is also downright mean with his property. Frogs do tend toward impassivity, particularly when they are bloated and have burrowed underground. So, when Kookaburra laughs, Tiddalik merely blinks. When Frill-Necked Lizard performs his antics, Tiddalik gives nothing away. When Brolga performs her extravagant dance, Tiddalik merely sits.[6] Miserly to the last, he is doubly mean in spirit; he will not share his resources, but neither will he feel the joy that comes from light-hearted performances. Only Eel can break the mould, doing so by performing something of a dance of death, writhing, and contorting in the way that eels are prone to do when caught or stranded (as in a drought). Eel's performance is not a happy one,[7] but it has the desired effect. Tiddalik cannot resist. It is as if an act of *Schadenfreude* is required to make him give up what he owns. In another variant of the story (Eggleston, 1964, pp. 70–77),[8] the creature who finally gets Tiddalik to burst out laughing is Worm, who performs the feat through tickling Tiddalik. In yet another version, it is Platypus who makes Tiddalik laugh merely by making an appearance, Tiddalik thinking Platypus's appearance as part-mammal and part-bird so utterly ridiculous that he can no longer control himself (Troughton, 1994). In all these variants, Tiddalik resists every attempt to make him share and enjoy. He laughs only when forced to (tickled) or in inappropriate circumstances (at other's problems or faults). He does not *share* his humor and he is not an agent of *community*. If the universe is imagined as moral in form, we might think that drought is characterized by the ethical failure of water itself—but in this case it is the moral failure of Tiddalik as the "master of water."

---

[6] I need hardly point out that the antics of Kookaburra, Frill-Necked Lizard, and Brolga are based on the customary behavior (calling, fleeing, and mating) of these creatures.

[7] In Mountford's version it is described as "grotesque" (Roberts and Mountford, 1973, p. 24). In Ellis's version (2004, p. 16) it is said that, because of the drought, Eel "had nowhere to live. He needed water desperately. He would soon die without it, and he was angry ... VERY angry indeed!"

[8] The tickling idea appears to have its origin in one particular version of the original Kurnai story (Curr, 1887, pp. 547–548).

It is often said that a principle of reciprocity is at the heart of Aboriginal sociality, which often now translates into the idea that Aboriginal people "care and share." It is important to understand exactly what this means. The idea of reciprocity was famously described by French anthropologist Marcel Mauss (1954) in formal terms which we might translate as "for every action there must be a reaction." As noted above, Rose suggests that this same formula is at the heart of Aboriginal cosmology, but in the specifically human domain the principle devolves into the idea that whatever is given should be returned, because a gift always carries with it something (a "spirit") which connects it to the giver and which demands that the receiver is also given a sense of being in debt. Some such idea is at the heart of all human transactions—not only narrowly "economic" transactions, but all exchanges of material goods, gestures, services, and communications which create social relations and judgments of value. In that sense, the elementary unit of society is not "the individual," because an individual is by definition a completely autonomous unit. Such persons do not exist in the real world, because life must be lived through relationships. Hence, the fundamental unit is not singular; it is *plural*—not "one" but "two." In Aboriginal societies, this fact is clearly articulated through systems of extended kinship and affinity. At the heart of each lies an overarching social dualism known to anthropologists as moiety organization.

Generally speaking, classical and neoclassical Aboriginal societies are divided into two halves—"moieties"—which must mutually articulate in the achievement of reproduction through time. The basic principles of moiety organization are very straightforward: a person is always assigned to only one moiety (sometimes the father's, sometimes the mother's), but one must always marry and have children with a partner from the moiety which is not his or her own. These basic rules, in tandem with many others, generate complex sets of relationships governed through a moral economy of demands and obligations—"sharing." Complicated as this moral economy might be, the basic notion of sharing is premised on the straightforward idea that one half of society always depends on the other half through systemic reciprocity. In some areas, the moieties do not bear names and people simply refer to themselves egocentrically as "us" and "them." In other areas, the moieties are named, as in the Yolngu case in northeastern Arnhem Land, where they are called *Dhuwa* and *Yirritja* (Warner, 1969, pp. 29–33; Morphy, 1991, pp. 43–45). In many cases, the moieties are totemically aligned.

In the Yolngu case, for example, phenomena in nature (totems) are assigned to either the *Dhuwa* or *Yirritja* moiety, achieving what Stanner (1989, p. 137), in another context, calls the correlativity of cosmos and society. In yet other areas, the moiety names are themselves totemic. In the western district of Victoria, for example, a number of groups once called their moieties Black Cockatoo and White Cockatoo (Radcliffe-Brown, 1951).[9] Elsewhere in southeastern Australia, the moieties have been called Eaglehawk and Crow (Blows, 1995).[10] In each case, the basic symbolism is that of "same but different," thus giving autonomy and relatedness equal weight in the definition of sociality. Black and white cockatoos, for example, are similar in that they are members of the same family Cacatuidae, yet they could not be more different in the coloring.[11] Eaglehawk and Crow are both carnivores, but while one is a superlative hunter, the other is a scavenger. While the lone Eaglehawk is a hunting bird *par excellence*, it often has its prey stolen by mobs of crows. Hence, Eaglehawk and Crow are also *competitors* and their relationship is marked by struggle and the will to win. Hence, harmony between Eaglehawk and Crow only obtains to the extent that their antagonistic qualities—Eaglehawk's strength and Crow's cunning—are balanced. So too it is with humans, since Aboriginal life is marked by "the struggle between circumstance and principle, identity and relation, independence and interdependence" (Stanner, 1989, p. 164). Aboriginal moral economies, which are also *political* economies, are harmonious only to the extent that there is a balance of both obligations *and* demands (Peterson, 1997), meaning that sharing may sometimes need to be forced. Such is the case with the demands made on Tiddalik.

The role of conflict and competition in Aboriginal moral economies is most clear toward the ending of the Tiddalik story. Pelican's negotiation of an ambiguously black–white identity resonates with the Black Cockatoo and White Cockatoo moiety symbolism of western Victoria. It also resonates with Eaglehawk and Crow moiety symbolism to the extent that it explicitly involves conflict, specifically over the proposed marriage between Pelican and the reluctant woman who wished to stay with her kin. But why does the Pelican have such a special place in relation to Tiddalik's flooding of the landscape? Part of the answer lies in the fact that, throughout arid and semiarid Australia, pelicans only make an appearance after heavy rains. While distributed over the entire range of the continent, as water birds which live entirely on fish and other aquatic creatures, they can only frequent large areas (the more arid ones) intermittently and opportunistically when dry water courses and lakes fill with water. Otherwise, their

---

[9]More specifically, according to Dawson (1888, p. 26), the long-billed corella (*Cacatua tenuirostris*) and the red-tailed black cockatoo (*Calyptorhynchus magnificus*) (Frith, 1976, pp. 250, 255).

[10]That is to say, wedge-tailed eagle (*Aquila audax*) and Australian raven (*Corvus coronoides*) (Frith, 1976, pp. 129, 578).

[11]Not simply because they are white and black, respectively, but because one has a splash of red in its tail while the other has a red splash on its head. The species also embody other contrasts which I do not discuss here.

nomadic movements are restricted to coastal zones and areas where rivers and lakes are permanent. Like frogs, then, pelicans indicate the presence of water, but the two are also opposites. While Tiddalik signifies the presence of water during a drought (and therefore also the *absence* of water), Pelican signals the presence of water during a flood. This more or less explains the small detail in the flood story which attributes the discovery of Tiddalik's possession of water to migratory or nomadic seabirds, among which Pelican would number. It does not, however, explain the fact that Pelican was originally black.

Pelican's exclusively black feathers are obviously a foil to the manner in which he is later to paint himself with white clay to signify embarking on an avenging expedition. In effect, Pelican's white paint transforms him into a warrior intent on killing the male relatives of the woman he wished to take as a wife (Fig. 2). His anger arises as the result of their woman's refusal, which is hence formally analogous to Tiddalik's refusal to supply water—although while Tiddalik is passive and has to be acted upon (tricked) before giving up his water, Pelican is the passive recipient of a ruse which means that he fails

Figure 2   Why the Pelican is black and white? This artwork was completed by Amy Mobourne, a young Kurnai woman, when she was a pupil at Bairnsdale Primary School in Central Gippsland. The originally published image can be found at http://www.iearn.org.au/fp/stories/754.htm. As a small gesture of repatriation of the Tiddalik story, Amy receives all royalties from this chapter. © Amy Mobourne, Bairnsdale, Victoria, Australia. (See Color Plate 13.)

to get what he both desires and is, as he understands it, owed by the men he has rescued. This too, then, is a moral failure, although the resulting imbalance pertains to warfare; that is, it is in a sociological rather than cosmological register. War is by definition "negative reciprocity" (Sahlins, 1974), in this case the cost of a lack of a woman being measured by the intended death of her male relatives. But Pelican fails in his intention. Through meeting one of his own kind (black), but appearing to his *alter ego* in an opposite guise (white), both of their "one-sided" lives are brought to an end—first because the black pelican kills the white pelican, and second because pelicans have been checkered ever since. The strange and aggressive mirroring leads to a synthesis, with pelicans coming to symbolize the dualistic motif of complementary opposition, which in turn represents the normal social state of reciprocity and lawful exchange embodied in moiety organization.[12] It is at this moment, when ordinary pelicans come to visually embody and reconcile contradictory social forces, that the world at large is returned to a normal state. Neither drought nor flood obtain in a landscape that is at once watered, sunlit, and free of conflict. Cosmology and society have, in Stanner's terms, become "correlative."

It is axiomatic in the study of myth, at least at the level of analysis employed here, that repeated occurrences of a particular motif, such as that of the flood, do not necessarily signal a common meaning of the motif across its varied manifestations. There are certain approaches to mythology (some psychoanalytic, some phenomenological) which suggest that recurrent motifs come to light in different cultural contexts because mythic imagery and symbolism is ultimately rooted in common features of the body and certain experiences and relationships typical of a universal human condition. This claim may well be true,[13] but the meanings of recurrent motifs cannot in themselves be completely reduced to any single property. Rather, they need to be fully contextualized and therefore relativized through time and space. As suggested earlier, a flood is by definition an excess of water. A flood can be destructive, but it can also be creative—and sometimes, as in the case of the Tiddalik story, it can be ambiguously poised between these two conditions as the bringer of both life and death. Human interest in such issues is undoubtedly universal and ultimately rooted in typically human reproductive needs. But these needs have been refracted through the long haul of

---

[12]This analysis is perhaps confirmed by another Kurnai story which makes it clear that *Borun* (Pelican), along with his wife *Tuk* (Musk Duck), was the original progenitor of the Kurnai people (Howitt, 1904, p. 485). It is essential to note, however, that the Kurnai were unusual insofar as explicitly labelled moieties were absent from their social structure, with their role partially being performed instead by gender totemism (Howitt, 1904, 148–149).

[13]And I have argued that it is true elsewhere (Morton, 1993).

human evolution and the development of localized societies, cultures, and ecologies. Of particular interest in this regard is the fact that the story of Tiddalik, in spite of having its origins in classical Aboriginal societies articulated by economies, social structures, and cosmologies, very different from those typifying "modern Australia," has become a tale of great interest to "modern Australians." Should we think that his continuing relevance is simply a function of his universal or "archetypal" appeal to human consciousness? Perhaps, but that is not the whole of the story, and for some purposes it may be the least interesting part.

I noted earlier that contemporary versions of the Tiddalik story are usually pitched to children. I also noted that contemporary versions tend to leave out the last part of the story concerning Pelican's move toward an embodied balance of black and white. Hence, in its contemporary guise the Tiddalik story has undergone something of a double reduction. Perhaps the final part of the story seems less relevant today and it is less easy for people to draw a lesson from it about social relations, given that its meaning is so closely linked to Indigenous social relationships which have obtained only under classical conditions and which are less frequently relevant today. On the other hand, the earlier section of the story perhaps has more universal appeal, particularly in an environmental context where droughts and floods are routinely reported and mythologized as aspects of an "extreme" continental and national landscape where Australians battle with or struggle to accommodate powerful natural forces. As this essay comes to a close, however, I want to argue that the compromise that Pelican represents through the reconciliation of black and white contains a message every bit as relevant to the contemporary context as the one contained in Tiddalik's deceptively simple classroom lessons about greed and cooperation.

III. CONCLUSIONS

While the initial part of the Tiddalik story may seem to be entirely cosmological in character, it embeds a sociological dimension. The dimension is specific to societies organized by moieties and formed by smaller extended families sometimes referred to as totemic clans. A continent-wide feature of such organization was (and sometimes still is) the periodic "national" gathering of totemic groups for large scale festivals, something which was characteristically achieved when resources were temporarily in abundance, usually after heavy rains. These festivals were a time when people should first settle old scores in order to achieve a harmonious collective identity through spectacular performances of totemic dramas—a time when people felt most free of conflict and most completely happy in spirit. The gathering of all the animals around Tiddalik, each willing to perform its characteristic behavior, is evidently analogous to such a festival. Their profound desire to laugh, be happy, and be free of conflict is threatened by Tiddalik's overarching meanness—"overarching" because Tiddalik alone could provide the product (water) most essential for a harmonious moral and political economy. Only when Tiddalik laughs can the process of integration get started, culminating in Pelican's achievement of balance.

This is not just a lesson about people sharing; it is a lesson about *institutional* balance. As Malinowski (1993) once pointed out, totemic clan organization represents the social form that a complex division of labor takes in classical Aboriginal societies, with each totemic group being responsible for the care of its country and the reproduction of the resources (including water) with which it is most intimately associated. Far from exclusively encompassing an "acentred land ethic" (Rose, 1998b, p. 296) in which human beings are simply embedded in the environment, Aboriginal worldviews were also anthropocentric enough to give rise to a notion of human control and stewardship. Moreover, the tendency to humanize the environment, to anthropomorphize it and see it totemically in terms of spiritual beings acting as humans might, is an adjunct to anthropocentrism, for it meant that living people could enter into contracts and rule-bound exchanges with the world at large—the Law. But as the Tiddalik story suggests, the Law is ultimately a collective human enterprise. In that sense, while a frog can be seen as an environmental regulator because of its objective characteristics, it only becomes an ethical regulator when, as in myth, it is engaged in a set of metaphors concerning specifically human relationships. In a contemporary context, when one hears much about the "ethics of water" in relation to the way this vital resource should be used and managed, Tiddalik's lesson is not simply that we need to be less greedy; it is that we need to arrange our institutions (the family, the economy, the government, science, religion, etc.) in a responsible and collectively appropriate manner so that human life can be sustained.

This begs a final word. Stanner has spoken of a "mood of assent" (1989, p. 56) in Aboriginal religion and part of this mood is captured by the way in which Aborigines have anchored the Law in the abiding features of the landscape and the wider cosmos. This does not mean that the Law did not encompass change. The Tiddalik story has its origins in a particular part of Australia, the area around Port Albert, to the east of Wilson's Promontory in Gippsland, Victoria. As Massola (1965, pp. 112–113) points out, the inlets and complex pattern of islands around Port Albert are the model for the flood found in the story, while the far flung White Rock, some 16 miles south of Port Albert, is the place which marks the continuing presence of Pelican (Fig. 3). One might reasonably assume, in-line with the argument put forward by Flood (1983, pp. 179–180), that the Tiddalik story preserves a memory of the time when this whole area was dry land prior to its flooding

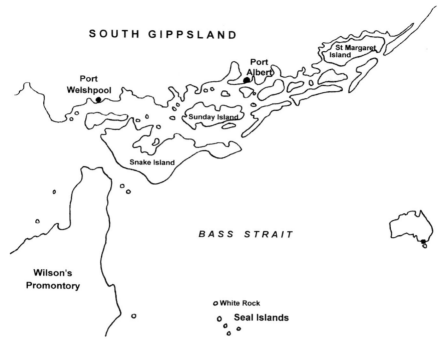

Figure 3 Map of the Port Albert Region illustrating the island landscape formed by Tiddalik's flood and the location of Pelican at White Rock.

in the Late Quaternary. If the assumption is correct, then it is evident that myth and history can be reconciled (Sutton, 1988), and Tiddalik's lesson in an era of global warming becomes yet more profound. Whether we avoid the coming flood or simply manage it, we will certainly require some balance in the agencies ultimately responsible for sustaining human life and Law.

ACKNOWLEDGMENTS

A number of people have given me assistance with this chapter, particularly in relation to the tracking of sources. While my thanks go to all of them, I particular wish to acknowledge help given by Julie Finlayson, Marcia Langton, Ray Madden, and Michael Pickering. I am also grateful to Lydia Fucsko and Amy Mobourne for granting permission to reproduce Figs. 1 and 2, respectively. Thanks also to Aldo Poiani for stopping me rambling on.

REFERENCES

Berndt, R.M. and Berndt, C.H. (1970) *Man, Land and Myth in North Australia: The Gunwinggu People*. Michigan State University Press, East Lansing.

Berndt, R.M. and Berndt, C.H. (1988) *The Speaking Land: Myth and Story in Aboriginal Australia*. Penguin, Melbourne.

Blows, M. (1995) *Eagle and Crow: An Exploration of an Australian Aboriginal Myth*. Garland Publishing, New York.

Cogger, H.G. (1979) *Reptiles and Amphibians of Australia*. A.H. and A.W. Reed, Sydney.

Curr, E.M. (1887) *The Australian Race: Its Origins, Languages, Customs, Place of Landing in Australia and the Routes by Which It Spread Itself over That Continent*, Vol. 4. J. Ferres, Melbourne.

Dawson, J. (1881) *Australian Aborigines: The Languages and Customs of Several Tribes of Aborigines in the Western District of Victoria, Australia*. George Robertson, Melbourne.

Eggleston, R. (1964) *When Yondi Pushed up the Sky*. Australian Publishing, Sydney.

Ellis, J. (2004) *More Stories from the Aboriginal Dreaming: A Selection from Many Areas Across Australia*. Kaliarna Productions, Penrith.

Flood, J. (1983) *Archaeology of the Dreamtime: The Story of Prehistoric Australia and Her People*. Collins, Sydney.

Frith, H.J. (Ed.) (1976) *Reader's Digest Complete Book of Australian Birds*. Reader's Digest, Sydney.

Henry, C.J. (1967) *Girroo Gurll, the First Surveyor*. W.R. Smith and Paterson, Brisbane.

Hornbuckle, J.W., Christen, E.W., Podger, G., White, R., Seaton, S., Perraud, J.-M. and Rahman, J. (2005) *Tiddalik: An Irrigation Area Model for Predicting and Managing Drainage Return Flows*. Australian National Committee on Irrigation and Drainage, Mildura.

Howitt, A.W. (1888) Further notes on Australian class systems. *J. Roy. Anthropol. Inst.* **18**, 30–70.

Howitt, A.W. (1904) *The Native Tribes of South East Australia*. Macmillan, London.

Jennings, J.N. (1971) Sea level changes and land links. In: *Aboriginal Man and Environment in Australia* (Ed. by D.J. Mulvaney and J. Golson), pp. 1–13. Australian National University Press, Canberra.

Jones, R. (1994) Pleistocene. In: *The Encyclopaedia of Aboriginal Australia* (Ed. by D. Horton), Vol. 2, pp. 876–877. Aboriginal Studies Press, Canberra.

Knudtson, P. and Suzuki, D. (1992) *Wisdom of the Elders*. Allen and Unwin, Sydney.

Lévi-Strauss, C. (1966) *The Savage Mind*. Weidenfeld and Nicolson, London.

Lévi-Strauss, C. (1968) *Structural Anthropology*. Penguin, Harmondsworth.

Lévi-Strauss, C. (1970) *The Raw and the Cooked: Introduction to a Science of Mythology I*. Jonathan Cape, London.

McConvell, P. and Thieberger, N. (2005) Languages past and present. In: *Macquarie Atlas of Indigenous Australia* (Ed. by B. Arthur and F. Morphy), pp. 78–87. Macquarie Library, Sydney.

Malinowski, B. (1926) *Myth in Primitive Psychology*. Kegan Paul, London.

Malinowski, B. (1993) The economic aspects of the *intichiuma* Ceremonies. In: *The Early Writings of Bronislaw Malinowski* (Ed. by R.J. Thornton and P. Skalnik), pp. 209–227. Cambridge University Press, Cambridge.

Massola, A. (1965) The Port Albert frog and the White Rock. *Vic. Nat.* **82**, 111–113.

Mauss, M. (1954) *The Gift: Forms and Functions of Exchange in Archaic Societies*. Cohen and West, London.

Morphy, H. (1991) *Ancestral Connections: Art and an Aboriginal System of Knowledge*. University of Chicago Press, Chicago.

Morton, J. (1993) Sensible beasts: Psychoanalysis, structuralism, and the analysis of myth. *Psych. Study Soc.* **18**, 317–341.

Mowaljarlai, D. and Malnic, J. (2001) *Yorro Yorro: Everything Standing up Alive*, (revised edition). Magabala Books, Broome.

Mudrooroo (1994) *Aboriginal Mythology: An A-Z Spanning the History of Aboriginal Mythology from the Earliest Legends to the Present day*. Aquarian, London.

Parker, K.L. (1953) In: *Australian Legendary Tales* (Ed. by H. Drake-Brockman, illustrated by Elizabeth Durack). Angus and Robertson, Sydney.

Peterson, N. (1997) Demand sharing: Sociobiology and the pressure for generosity among foragers. In: *Scholar and Sceptic: Australian Aboriginal Studies in Honour of L.R. Hiatt* (Ed. by F. Merlan, J. Morton and A. Rumsey), pp. 171–190. Aboriginal Studies Press, Canberra.

Radcliffe-Brown, A.R. (1951) The comparative method in social anthropology. *J. Roy. Anthropol. Inst.* **81**, 15–22.

Reed, A.W. and Hart, R. (1965) *Myths and Legends of Australia*. A.H. and A.W. Reed, Sydney.

Roberts, A. and Mountford, C.P. (1973) *The Dreamtime Book*. Rigby, Adelaide.

Roennfeldt, R. (1981) *Tiddalick, the Frog Who Caused a Flood: An Adaptation of an Aboriginal Dreamtime Legend*. Penguin, Melbourne.

Rose, D.B. (1998a) Consciousness and responsibility in an Australian aboriginal religion. In: *Traditional Aboriginal Society* (Ed. by W.H. Edwards), 2nd ed., pp. 239–251. Macmillan Education, Melbourne.

Rose, D.B. (1998b) Exploring an Aboriginal land ethic. In: *Traditional Aboriginal Society* (Ed. by W.H. Edwards), 2nd ed., pp. 288–296. Macmillan Education, Melbourne.

Sahlins, M. (1974) *Stone Age Economics*. Tavistock, London.

Smyth, R.B. (1878) *The Aborigines of Victoria: With Notes Relating to the Habits of the Natives of Other Parts of Australia and Tasmania*, Vol. 2. John Ferres, Melbourne.

Stanner, W.E.H. (1979) *White Man Got No Dreaming: Essays 1938–1973*. Australian National University Press, Canberra.

Stanner, W.E.H. (1989) *On Aboriginal Religion*. (Facsimile Edition) University of Sydney, Sydney.

Steele, J.G. (1983) *Aboriginal Pathways in Southeast Queensland and the Richmond River*. University of Queensland Press, Brisbane.

Strehlow, C. (1907) *Die Aranda- und Loritja-Stämme in Zentral Australien I: Mythen, Sagen und Märchen des Aranda-Stammes*. Joseph Baer, Frankfurt-am-Main.

Sutton, P. (1988) Myth as history, history as myth. In: *Being Black: Aboriginal Cultures in 'Settled' Australia* (Ed. by I. Keen), pp. 251–268. Aboriginal Studies Press, Canberra.

Thomas, W.J. (1939) *Some Myths and Legends of the Australian Aborigines*. Whitcombe and Tombs, Sydney.

Timaepatua, M.A. and Gumudul, R. (1977) Kwork Kwork the green frog. *In: Kwork Kwork the Green Frog and other Tales from the Spirit Time*, pp. 2–3. The Australian National University Press, Canberra.

Troughton, J. (1994) *What Made Tiddalik Laugh: A Lively Legend from Australia*. Penguin, London.

Turkle, S. and Papert, S. (1990) Epistemological pluralism: Styles and voices within the computer culture. *Signs* **16**, 128–157.

Unaipon, D. (2001) In: *Legendary Tales of the Australian Aborigines* (Ed. by S. Muecke and A. Shoemaker). The Miegunyah Press, Melbourne.

Vanderwal, R. (Ed.) (1994) *Victorian Aborigines: John Bulmer's Recollections 1855–1908*: (Compiled by Alastair Campbell) Museum of Victoria, Melbourne.

Warner, W.L. (1969) *A Black Civilization: A Social Study of an Australian Tribe*, (revised edition). Peter Smith, Gloucester, MA.

Wolfe, P. (1991) On being woken up: The Dreamtime in anthropology and in Australian settler culture. *Comp. Stud. Soc. Hist.* **33**, 197–224.

Understanding the Social Impacts of Floods in Southeastern Australia

CATHERINE ALLAN, ALLAN CURTIS AND NICKI MAZUR

SUMMARY

"Flooding" is as much a social construction as a physical reality. In this chapter, we examine three flood events in southeastern Australia, using social impact assessment and risk management as analytical frameworks. The lessons are that winners and losers from flood events and flood management activities can, and should, be predicted. Once vulnerable individuals and communities are identified, there are practical steps which can be taken to increase community resilience and capacity. Collaborative and participatory activities are fundamental to this resilience building, so current technically focused flood management planning may need to become more inclusive.

I. INTRODUCTION

Human memory, myth, and history abound with floods and flooding, and we know that floods will continue to be part of our lives. Indeed, there is

ADVANCES IN ECOLOGICAL RESEARCH VOL. 39   0065-2504/06 $35.00
                      DOI: 10.1016/S0065-2504(06)39009-5

evidence to suggest an increased incidence of flooding in the future as changes to our planet's climate alter the frequency of extreme weather events (Bronstert, 2003) and sea levels rise in response to the melting of ice sheets (Munk, 2003). Flooding involves inundation of land with water, but the concept of a "flood" is as much social as it is physical; it is the impacts on humans and their interests that define water covering land as a flood (Howarth, 2003).

The activities that societies undertake to manage their natural resources are largely shaped by the values they attach to those assets and their perceptions of the risks and rewards of interventions (see chapter by John Freebairn). Until the last half of the twentieth century, natural resource/ environmental management appeared to be fairly straightforward, as managers' major aims were the exploitation of natural resources and the protection of human assets. Now there is a bewildering range of values to be considered, including those underpinned by improved understanding of ecosystem complexity (Holling, 1995). For instance, regular flooding has been recognized as a boon for some delta farmers since agriculture began, but now flooding is more widely credited with a variety of ecosystem replenishment roles (see chapters by Sam Lake, Nick Bond, and Paul Reich and by Andrea Ballinger and Ralph MacNally in this volume). Our increasing understanding of social variety and cultural needs is also increasingly influencing natural management. To the extent that social and economic impacts of floods were considered in the past, the focus was mostly on crisis planning and management, disruption, the economic impacts of floods, and on economic recovery from them (Dynes, 2003). However, the management of natural resources generates costs, benefits, and other social impacts that are unevenly distributed, in space and time, across different individuals, groups, organizations, industries, regions, and cultures (Dale et al., 2001). Understanding those impacts is where social research can make an important contribution to floods and flood management.

In this chapter, we draw heavily on the theory of risk perception and social impact assessment (SIA) (Burdge, 1994; Coakes, 1999) to explore what we think are the principal social dimensions of floods. Social dimensions include, but are not restricted to, negative impacts, such as loss of life, livelihood, property, or recreation opportunity, and positive impacts derived from floodplain rejuvenation, including increased agricultural production and tourism. To ground our discussion, we have drawn on Australian case studies where we have direct personal experience. Allan Curtis grew up on the Snowy River in the East Gippsland region of Victoria before attending University in Melbourne and moving to northeast Victoria to live at Tallangatta and then Albury/Wodonga. In the 1990s, he was a ministerial appointee to the Murray–Darling Basin Ministerial Council Community Advisory Committee (MDB CAC) and was a member of the Reference Panel that explored

Plate 14

options for changing the operation of Hume and Dartmouth Dams. Catherine Allan was raised in northeast Victoria and also attended Melbourne University. She returned to work in her home town of Benalla where her property was inundated by the 1993 Broken River flood, before moving to Albury-Wodonga in time to catch the 1996 flood of the Murray and associated rivers.

II. SOCIAL IMPACT ASSESSMENT

While SIA is predominantly associated with understanding the impacts of proposed developments (which may include flood management structures), the process can also be useful for anticipating the impacts of events such as floods. Using SIA can improve our understanding of which social groups may be affected by floods/planned flood management activities and how any negative effects may be minimized and positive effects maximized. Hazard management can be seen as balancing the often conflicting objectives of hazard reduction, economic use, and environmental amenity of a hazardous area (Handmer, 1996). Flood hazard management and flood impact response decisions will (must) continue to be made, and those decisions will continue to create winners and losers. The complexity and subtlety of decision making for multiple outcomes, combined with greater understanding of the complexity of ecosystem responses, suggest that there will be always conflicting views and opinions. SIA provides a framework for working through the complexity and conflict to achieve equitable outcomes by assessing and predicting which members of society will be impacted by those decisions, and in what ways.

SIA explores how particular events or policies will affect people's way of life, their culture, and their community (Burdge, 1994; Bronstein and Vanclay, 1995). SIA may draw on economic assessments, but emphasizes the nonmonetary effects of an intervention. SIA uses a range of social science disciplines to anticipate the consequences of proposed actions or possible events. Among other things, SIA examines the unequal distribution of benefits and costs; changes in power structures; implications for family life, health and education; and effects on community cohesion and local organizations. SIA provides policy makers with a process for identifying and working through issues with stakeholders (Coakes, 1999).

SIA typically involves a number of steps, including:

Engaging the public: Identify affected individuals and groups and develop a process for engagement.
Identifying alternatives: Describe the proposed action/policy intervention and reasonable alternatives.
Profiling the social environment: Describe the baseline social environment.

Scoping impacts: Identify all potential impacts both positive and negative, including consideration of "winners" and "losers" and the extent change will be beyond that which might otherwise occur.

Evaluating impacts: Determine probability of occurrence, assess distributional impacts across individuals and groups, determine priority of issues to stakeholders, and suggest how individuals and groups are likely to respond.

Identifying alternatives and mitigation strategies: Canvas and prioritize strategies to mitigate undesirable impacts.

Monitoring: Develop a monitoring program that will flag deviations from proposed course and indicators of any unanticipated impacts.

An important aspect of SIA is the identification of social groups which may be impacted in both negative and positive ways (the winners and the losers), in particular in relation to individual and community well-being. The Australian Collaboration (2001, cited in Pritchard *et al.*, 2003, p. 13) suggests that social well-being has four major components:

1. Meeting basic human needs (including employment, education, adequate housing, and health)
2. Addressing inequality and poverty
3. Reconciliation and tolerance of others
4. Building social capital

Flooding or flood management activities may produce winners and losers by increasing or reducing any of these components of individual or group well-being.

For this chapter, we will concentrate on the last four steps of SIA: scoping, evaluation, mitigation, and monitoring. We will selectively draw on our case study experience to explore the social impacts of floods and identify lessons learned from a social research perspective, including about how our understanding of risk perception can lead to improved engagement with stakeholders affected by floods.

A. Case Study One: Winners and Losers and the Impact of the Snowy Scheme in East Gippsland

Work on the Snowy Mountains Scheme commenced in 1949. This massive system of dams and tunnels diverted water inland into the Murray and Murrumbidgee Rivers from the Snowy River that flowed from the Alps to the sea at Marlo near Orbost in East Gippsland (Victoria). The diversions were used to generate electricity and to support irrigated agriculture. Until very recently, general consensus held the Snowy Scheme as a wonderful example of nation building, and that most of those affected by it were winners. Clearly the inland irrigators benefited by having access to cheap and plentiful supplies of fresh water, as did anyone supplied with electricity generated with little pollution or emission of greenhouse gases. The farmers and townspeople on the Snowy floodplain also expected to benefit from fewer and less intense floods; previously flooding had been an annual event which had significant negative impacts on farm incomes, infrastructure and frequently resulted in losses of stock and people's lives.

By the early 1970s, it was apparent that districts formerly fed by the waters of the Snowy River were in serious economic decline. East Gippsland has a narrow economic base, being largely dependent on primary industries, including agriculture and timber harvesting from native forests. Despite the claimed national benefits of the Snowy Scheme, residents in East Gippsland began to identify a range of negative impacts that they thought outweighed any gains. Apart from the negative impacts on the ecosystems of the Snowy River and its estuary, including silting of the river and frequent closing of the estuary mouth, reduced flooding interrupted the replenishment of nutrients on the floodplain, which lowered farm output and increased production costs. With reduced freshwater flows, salt water incursions via the estuary mouth were thought to have reduced pasture production. Rising salinity levels in the river also prevented farmers from supplementing rainfall with irrigation water.

With the move away from harvesting Australia's native forests for timber, the East Gippsland timber industry was in serious decline throughout the 1980s and 1990s. The construction of a pulp mill in the district to utilize woodchips was seen as a potential growth catalyst. Locals were dismayed to find that the proposed mill was not viable, partly because there was insufficient fresh water flowing through the Snowy River. At the same time, dairy farmers and those cropping beans and corn faced increased competition from lower cost producers in the irrigation districts of the Murray and Murrumbidgee Valleys using water sourced, at least in part, from the Snowy River. Although the dairy industry continues in the Orbost district, there is now little cropping of beans or corn.

The irony of these examples has not been lost on the people living in East Gippsland, and they lobbied governments to acknowledge and address their concerns. With a small population, declining economic base and a history of support for conservative politicians, this isolated regional community had little power or influence and their pleas for assistance were largely ignored by state and federal governments. This all changed when Craig Ingram was elected to state Parliament in 1999, becoming one of a small group of independents holding the balance of power in the state of Victoria following the defeat of the Kennet government. Ingram was committed to restoring environmental flows in the Snowy River and agreed to support the Bracks

Labor government of Victoria that succeeded Kennet. In turn, the Bracks government agreed to secure commitments from the New South Wales state and Australian federal governments to return substantial flows of water to the Snowy River.

This little story highlights that the (human) losers from the operation of the Snowy River Scheme were only identified after major problems emerged, and action to manage some of these problems was only taken because a politician was willing to champion a cause and was in a political situation which allowed him to influence the state government.

B. Case Study Two: The Hume and Dartmouth Dams and How Understanding the Social Leads to Better Outcomes

Large water storages can (among other things) reduce the incidence and extremity of flood events in a river downstream of the dam, by inducing long-term flooding upstream of the wall. We include a brief account of the Hume and Dartmouth Dams here to highlight the complexity of managing water and flooding in a large and interdependent system.

Discussions about a Murray River water storage began in 1902, but agreement on the Hume Dam site did not occur until 1919. The Depression of 1929–1931 slowed construction, and a smaller dam than that originally planned was completed late in 1936. In 1952, most of the town of Tallangatta (Victoria) was moved to a site 8 km to the west to allow for the expansion of the dam's capacity. The Hume Dam supplies hydroelectricity and water to towns, industries, and irrigation properties downstream of Albury (New South Wales). Once again the immediate "winners" are easy to pick; irrigators, all electricity consumers, and the tourism industry in towns, such as Mulwala (New South Wales) and Yarrawonga (Victoria), which benefit from consistently high water levels. But there were immediate losers too, most obviously some of the residents of "old" Tallangatta and the farmers who had their land compulsorily acquired. As this open letter, written just prior to the flooding of Tallangatta by W.H. Ferguson, illustrates social impacts extend beyond economic impacts and include the loss of identity and attachment to place:

> *Perhaps I should apologise to the public for broadcasting the small affairs of a place unknown to many thousands living in Victoria, but this "history" is somewhat of an obituary notice, the poor little place is to be drowned, a very beautiful and fertile parish is to be drowned by the Hume Weir Waters. There are two ways of getting rid of people—fire and water. The Governments in their mercy have adopted water, we are to be merely drowned out; we should be grateful. We are not. We are told for every one*

of us driven off the land, ten others will be placed upon it, and will live happily ever afterwards. We grudge no people their happiness, but we dread our removal. Ministers of the Crown put the matter with brutal candour. "We won't pay for sentiment", saith one; another, "You cannot make omelettes without breaking eggs". We are the eggs, and though we are being enforced off the earth we should not squeal. The boasted policy of the governments is to settle the people on the land, they are unsettling us off the land .... The pioneers of our valleys fought nature; they carved smiling farms out of forests and swamps. They have left their sons a goodly heritage. We must go, but we go with sorrow ....

Cited in Swan (1987)

Thirty years later residents of the Tallangatta Shire were once again, in their words, "Dammed for Australia" (Swan, 1987). As part of efforts to "drought-proof" the Murray system, the decision was taken to construct Dartmouth Dam on the Mitta River above Lake Hume. This was a massive undertaking, the wall itself is around 400-m high and the dam capacity is larger than Hume. Construction was completed in the late 1970s. Apart from the obvious economic and social impacts that resulted from a huge investment and the influx of workers and their families in a small rural community, there have been other longer-term impacts, to which we will return. The idea that the Murray system could be "drought-proofed" might appear arrogant and somewhat dated, but it highlights the extent to which we are prepared to intervene to manage perceived risks, in this case drought. Fundamental to understanding the possible impacts of an event is an understanding of risk and perceived risk. It is important to explore the concept of risk perception in the context of flooding, and in our case, from a social research perspective.

While "risk" is a term used daily, its more conventional, technical meaning is used to refer to "A combination of the probability, or frequency, of occurrence of a defined hazard and the magnitude of the consequences of the occurrence: how often is a particular potentially harmful event going to occur, [and] what are the consequences of this occurrence?" (Harding, 1998, p. 167). This definition illustrates how in recent years, the notion of "risk" is increasingly focused on possible unwanted outcomes, moving away from a traditional sense of risk taking as a potentially positive activity. Furthermore, terms, such as "risk management" and "acceptable levels of protection," assume a degree of understanding of the concept of risk, acceptance of how it is measured and some level of consensus on how it should be managed.

The reality is that understandings of, and responses to, risk are highly varied, because risk is socially constructed. That is, "risk" does not exist "out there," independent of human minds and cultures, waiting to be measured. Instead, humans have invented the concept of risk to help them understand and cope with the dangers and uncertainties of life (Slovic, 1999). All people, irrespective of their role in society, use speculative frameworks to make sense of the world and make selective judgments in their responses to different kinds of risk. These so-called "nonrational" factors are not necessarily incorrect, but there is likely to be significant variation between people in their understandings and responses.

Since risk is socially constructed, the perception of risk will vary with the particular characteristics of the risk (Merkhofer, 1987; Finucane, 2000), the psychological aspects of individual and societal decision making (Covello et al., 1984; Finucane, 2000), and sociopolitical factors (Slovic, 1999). Differences between the way "experts" and the "lay public" perceive risk often result in situations where "experts" use increasing amounts of science-based evidence to "convince" the public that the risk is either negligible and/or under control, although there is substantial evidence to show that this approach actually invites public suspicion (Petts and Leach, 2000; Randall, 2002). Citing Jasanoff (1998), Brown and Damery (2002) note that the primacy of scientific knowledge and its attendant values of rationality and objectivity have led to the belief that we *can* (and should) distinguish between "real" risks (as seen by experts) and risks viewed by lay audiences. Within this way of thinking the latter, "perceived risk," is simply distorted reality that is informed by ignorance. This persistent and dominating belief has resulted in risk management policies that focus on communicating with "irrational" publics to remove the perception of risk, while reducing any "real" risks with technological solutions (Brown and Damery, 2002).

Continued reliance on this type of risk management and communication presents hazards of its own for decision makers, namely lowering public trust in flood planning and management. As mentioned, sociopolitical contexts are crucial to influencing risk perceptions. That is, the degree of trust in institutions, communicators, and particular messages is an important influence on how risk is perceived and will influence how people respond to communications about (flood) risks (Petts and Leach, 2000). People are concerned about the fairness of decision processes and outcomes (Beckwith et al., 1999). Where institutions and decision makers are perceived to be credible (believable, trustworthy, reliable, competent, objective, consistent, transparent), there is likely to be lower perceived risk and more positive responses to communication about that particular risk (Petts and Leach, 2000; Trettin and Musham, 2000).

Our experience with the Hume and Dartmouth Dams provides some useful illustrations of how organizations attempting to engage the public on topics related to floods and river management can achieve better outcomes if they understand that risk perception is socially constructed and they use timely and meaningful community engagement practices (Aslin and Brown, 2002). The Mitta Valley in Victoria has relatively high rainfall, fertile soils, and cool summer temperatures that make it ideal dairy country. In this environment, Mitta farmers have had little need to irrigate their pastures. During the 7 or 8 years it took to fill Dartmouth, the Mitta River below the dam wall was reduced to a trickle. Soon after the fill phase for Dartmouth commenced, Mitta farmers began to report declining pasture and milk production. They attributed their production losses to lower groundwater tables as a result of reduced river flows and, in particular, to the elimination of winter/spring flooding that had occurred most years prior to dam construction. By the mid-1980s, farmers were so concerned that they began to seek access to irrigation water to maintain pasture production, but with little success. Governments did not believe their claims of lost production, and the Mitta farmers had less political influence than the highly organized downstream irrigators who claimed that the water in Dartmouth belonged to them, not the farmers in the Mitta Valley.

With Dartmouth Dam operational, the Murray–Darling Basin Commission (MDBC) had increased confidence that there would be sufficient water to refill Hume Dam each year and reduce the risk of insufficient water for the start of each irrigation season. The MDBC was then able to adopt a less conservative approach to the management of Hume Dam. Among other things, this meant that more water could be released for sale to downstream irrigators each year. Again there were winners and losers from this change in management. The losers included a small group of landholders on the floodplain near Albury who had parts of their property flooded each summer as the MDBC attempted to move large volumes of water through the river and its anabranches.

In early 1997, in response to lobbying by the Mitta and Albury farmers, the MDBC established an independent stakeholder Reference Panel to review the operations of the Hume and Dartmouth dams (MDBC, 1999). The intention was to explore ways that a whole-of-system approach could lead to improved environmental outcomes. This was an ambitious undertaking. Management of the Murray system is complex and the Reference Panel included stakeholders with opposing views and powerful connections. The MDBC appointed a highly skilled and respected chair and adopted sound community engagement practices, including ensuring that all major stakeholders were represented. When stakeholders had little organizational support, additional resources were provided to ensure that full representation was more than simply a token gesture. Importantly, the Reference Panel's level of authority or decision-making power was clearly explained at the start, and sufficient time was allocated for familiarizing stakeholders with different parts of the system and their communities. A substantial investment was made to identify the impacts of proposed changes to dam operation, and processes were adopted that allowed participants to develop an understanding of other stakeholder positions, including the trade-offs associated with any changes in dam operation. As the literature suggests, the quality of the processes adopted, or the attention to procedural justice, is as important as the outcomes of this type of engagement process (Syme and Eaton, 1989; Lawrence *et al.*, 1997; Lauber and Knuth, 1998; Grant and Curtis, 2004).

Among other things, the Reference Panel employed a consultant to assess the economic impacts of reduced flooding and lower groundwater levels on agricultural production in the Mitta Valley. The consultant's report confirmed the farmers' concerns about the risks and negative impacts and resulted in payments of several million dollars in compensation to Mitta farmers. Another, perhaps less immediate outcome of this process was that when the MDBC was forced to release large volumes of water from Hume Dam in 1996 after the dam wall moved and threatened to collapse, farmers below Hume Dam, who had been involved or represented on the Reference Panel, accepted this action and the MDBC offered them compensation for the impacts of that flooding on their crops, fences, and other infrastructure.

C. Case Study Three: Two Victorian Flash Floods of the 1990s; How a Flood Event Is Experienced Depends on the Vulnerability and Capacity for Recovery of the Community Involved

In October 1993, the Northeast Victorian town of Benalla was flooded after more than 200 mm of overnight rain swelled the Broken River (Keating, 2005; see also Fig. 1). Residents who lived near the river, and who were awake at dawn, saw the floodwaters rapidly rise over the river banks and move across gardens and streets, meeting the surging waters from flooded storm-water drains. By early morning much of the central area of the Benalla township, including the central business district, was under 1–2 m of muddy, swirling water. In the street where Catherine Allan lived some newer houses on concrete slabs were flooded, while older houses on stumps sailed serenely above the waters, although the floor level on some of these perched just at the surface, so that the wash from vehicles (including the four-wheel drives of sightseers) brought the waters lapping in. The floodwater remained for several days, then slowly receded, leaving behind a patchwork of damaged and unaffected houses, a thick carpet of mud over all the gardens and streets of the impacted area, and an all-pervading stench. Keating (2005) describes the rapid and effective community responses to the shock of this rapid flooding and its aftermath. Activities included a large public meeting outside the municipal offices, hundreds of small neighborhood meetings, and acts of support for the townspeople from farmers and their farming industry

Figure 1 The aptly named Riverview Road, Benalla, Victoria, October 1993 (photo by C. Allan). (See Color Plate 14.)

groups. The Benalla Recovery Committee, which morphed into the Benalla Flood Action Committee, gained some immediate relief for townsfolk, and continued to work for many years to gain further financial support for residents and businesses negatively impacted by the flooding. In short, Benalla residents were stunned, and many severely impacted by the flood, but the Benalla community had a degree of financial and social resilience which enabled it to immediately strive to return to normal business and other life.

In June 1998, the Tambo Valley in eastern Victoria also experienced a flash flood which damaged infrastructure and physically isolated many in the community. However, the physical impacts were only part of the flooding story. Monson (2004) documents that the torrential rains and gale force winds which caused the Tambo floods were preceded by several years of drought, low prices for wool, timber mill downsizing, rationalization of government services, and an outbreak of Ovine Johne's Disease that combined to leave the Tambo Valley vulnerable because of a limited capacity to respond to further hardship. When it hit so suddenly, the flood was simply the final straw for many in the community. Monson observed that residents felt a loss of control and a sense of failure. Assistance from outside the community was needed, and came from the Victorian government, which adopted a holistic approach to addressing the impacts of the flood itself, and some of the conditions which had left the community in such a vulnerable state. Then Premier, Jeff Kennet, described the East Gippsland flood recovery as "one of the most effective he has seen after a natural disaster anywhere in Victoria over the 22 years of his parliamentary career" (Office of the Chairman of the East Gippsland Taskforce, 1998). Included in this response was a $4-million "environmental package," which funded programs related to pastures, vegetation, and land management. Through community meetings, staff from a number of state agencies, local government, and community groups used their skills and government funds to negotiate restructuring of property ownerships (more viable holdings) and changed land use practices.

The participatory, community approach taken to respond to the Benalla and Tambo Valley Floods is in line with much of current thinking on flood hazard management. For example, Kweit and Kweit (2004) recommend that flood management authorities move beyond conventional participation mechanisms and seek to mobilize and provide citizens with meaningful opportunities for input, and Few (2003) argues for greater community involvement in intervention programs.

These two physical events were similar in being flash floods in rural townships. However, Benalla people were able to wrench back control of their lives by drawing on the numerous reserves (financial and social) of the town, while the Tambo Valley community had less of these reserves to draw on because of historical events. These two examples raise the idea of social "vulnerability"; a further, large-scale example also comes from the 2005 tragedy of Hurricane Katrina in New Orelans, United States. The scale of the disaster in New Orleans was clearly exacerbated by it impacting on many communities with reduced capacity to avoid the event; for instance, economic hardship left people with few transport options, and many had to stay put and face the storm and flood.

Vulnerability is a term used in many different ways, but conventional use equates it with susceptibility to impacts (Few, 2003). Implicit in this usage is that there is variation in individual's and community's capacity to cope with events. Coping capacity is increasingly seen as a key component of households'/communities' level of vulnerability. It has social, political, institutional, and cultural dimensions—households/communities have access to different types and levels of resources to help them cope (Few, 2003).

III. LESSONS

From these examples, and by thinking within an SIA framework, we can draw a number of lessons that could assist those charged with flood hazard management.

1. Winners and losers from flood events and flood management activities can be (and should be) predicted. This prediction can be improved by:
   a. Including a range of people in the projection of risks
   b. Taking risk perception into account rather than relying on educating it out of people
   c. Taking a wide range of values into account
2. Vulnerable individuals and communities can be predicted
   a. Although vulnerability is scale dependent and dynamic, retrospective analysis of the impact of floods on communities can reveal indicators for vulnerability
   b. These indicators could be helpful when assessing other districts at other times
3. Because of 1 and 2, steps can be taken to minimize harm from flood events by:
   a. Reducing vulnerability by increasing community resilience/capacity
   b. Involving communities in both the prediction and management of the impacts of floods

There is no such thing as a "no risk" society, so the mitigation of negative impacts on the vulnerable and poor in our society needs to be a major part of flood impact planning. This requires some shift from the current, almost exclusive, focus on the exciting event of the physical flooding, and admiration for communities as they immediately "pull together" to overcome adversity. While these invigorating and uplifting stories remain important, an effort must be made to see beyond these to address the complex issues we have raised in this chapter. There is actually some difficult reflection and serious planning required. The link between land use planning and flood management clearly needs strengthening, and we suggest this should be participatory, inclusive of planning. Participatory planning facilitates active learning from past events and current knowledge, and fosters consideration of the full range of consequences from future activities and events for diverse actors (Curtis *et al.*, 1995; Schusler *et al.*, 2003). There are three major benefits which result from following a genuinely inclusive and participatory approach:

1. Many negative impacts can be identified in advance, and mitigated
2. Increased knowledge and understanding of risks and perceived risks as local and nonexpert knowledge is pooled with expert knowledge
3. Communication of risk will be improved

IV. CONCLUSIONS

We have shown that SIA can be used to understand some social aspects of floods and flooding and, because of its inherent participatory approach, SIA can also provide a means of empowering communities that may be impacted by floods. The realization that flooding can be considered to be the result of social, as well as physical, processes creates opportunities for flood hazard management to be better tuned to the needs of the communities in which hazard management occurs.

ACKNOWLEDGMENT

We thank John Handmer for his comments on the draft of this chapter.

REFERENCES

Aslin, H.J. and Brown, V.A. (2002) "Research Report for Good Practice Community Engagement for the Murray-Darling Basin." Bureau of Rural Sciences, Canberra.

Beckwith, J.A., Colgan, P.A. and Syme, G.J. (1999) Seeing risk through other eyes. In: *Contaminated Site Remediation: Challenges Posed by Urban and Industrial Contaminants* (Ed. by C.D. Johnston), pp. 49–56. Proceedings of the 1999 Contaminated Site Remediation Conference, Centre for Groundwater Studies, CSIRO Land & Water, Wembley, Western Australia.

Bronstein, D. and Vanclay, F. (Eds.) (1995) *Environmental and Social Impact Assessment*. John Wiley, New York, USA.

Bronstert, A. (2003) Floods and climate change. *Risk Anal.* **23**, 545–557.

Brown, J.D. and Damery, S.L. (2002) Managing flood risk in the UK: Towards an integration of social and technical perspectives. *Trans. Inst. Br. Geogr.* **27**, 412–426.

Burdge, R.J. (Ed.) (1994) *A Conceptual Approach to Social Impact Assessment: Collection of Writings by Rablel J. Burdge and Colleagues*. Social Ecology Press, Middleton, Wisconsin, USA.

Coakes, S. (1999) *Social Impact Assessment: A Policy Maker's Guide to Developing Social Impact Assessment Programs*. Bureau of Rural Sciences, Canberra, Australia.

Covello, V.T., Winterfeldt, D. and Slovic, P. (1984) Communicating scientific information about health and environmental risks: Problems and opportunities from a social and behavioral perspective. In: *Uncertainty in Risk Assessment, Risk Management and Decision Making* (Ed. by V.T. Covello, L.B. Lave, A. Moghissi and V.R.R. Uppuluri), pp. 221–239. Plennum Press, New York.

Curtis, A., Birckhead, J. and De Lacy, T. (1995) Community participation in landcare policy in Australia: The Victorian experience with regional landcare plans. *Soc. Nat. Resour.* **8**, 415–430.

Dale, A., Taylor, N. and Lane, M. (2001) *Social Assessment in Natural Resource Management Institutions*. CSIRO Publishing, Collingwood, Victoria.

Dynes, R.R. (2003) Noah and disaster planning: The cultural significance of the flood story. *J. Cont. Crisis Manage.* **11**, 70–177.

Few, R. (2003) Flooding, vulnerability and coping strategies: Local responses to a global threat. *Prog. Dev. Stud.* **3**, 43–58.

Finucane, M.L. (2000) "Improving Quarantine Risk Communication: Understanding Public Risk Perceptions (Report No. 00-7)." Decision Research, Eugene, Oregon.

Grant, A. and Curtis, A. (2004) Refining evaluation criteria for public participation using stakeholder perspectives of process and outcomes. *Rural Soc.* **14**, 142–163.

Handmer, J. (1996) Policy design and local attributes for flood hazard management. *J. Cont. Crisis Manage.* **4**, 189–197.

Harding, R. (1998) *Environmental Decision-Making: The Role of Scientists, Engineers and the Public.* The Federation Press, Leichhardt, NSW.

Holling, C.S. (1995) What barriers? What bridges? In: *Barriers and Bridges to the Renewal of Ecosystems and Institutions* (Ed. by L.H. Gunderson, C.S. Holling and S.S. Light), pp. 3–34. Columbia University Press, New York.

Howarth, W. (2003) Private and public roles in flood defence. *Non-State Act. Int. Law* **3**, 1–21.

Jasanoff, S. (1998) The political science of risk perception. *Reliab. Eng. Syst. Saf.* **59**, 91–99.

Keating, T. (2005) After the deluge—some reflections on more than twenty years of disaster management. Fourth Victorian Flood Management Conference, September 2005, Shepparton, Victoria.

Kweit, M.G. and Kweit, R.W. (2004) Citizen participation and citizen evaluation in disaster recovery. *Am. Rev. Publ. Admin.* **34**, 354–373.

Lauber, T.B. and Knuth, B.A. (1998) Refining our vision of citizen participation: Lessons from a moose reintroduction proposal. *Soc. Nat. Resour.* **11**, 411–424.

Lawrence, R.L., Daniels, S.E. and Stankey, G.H. (1997) Procedural justice and public involvement in natural resource decision-making. *Soc. Nat. Resour.* **10**, 577–589.

Merkhofer, M.W. (1987) *Decision Science and Social Risk Management.* D Reidel Publishing Company, Dordecht.

Monson, R. (2004) The 1998 floods in the Tambo Valley. *Int. J. Mass Emerg. Disasters* **22**, 61–68.

Munk, W. (2003) Ocean freshening, sea level rising. *Science* **300**, 2041–2043.

Murray–Darling Basin Commission (MDBC) (1999) "Hume and Dartmouth Dams Operations Review Reference Panel: Final Report and Recommendations." Murray–Darling Basin Commission, Canberra, Australia.

Office of the Chairman of the East Gippsland Taskforce (1998) "East Gippsland Flood Recovery 'A Wonderful Partnership'" <http://www.dpc.vic.gov.au/domino/Web_Notes/pressrel.nsf/0f717473d3a42d884a2566710008e993/0152c26b5 2603334a25667800097e94OpenDocument&Click = >.

Petts, J. and Leach, B. (2000) "Evaluating Methods for Public Participation." R&D Technical Report E135. Environment Agency, Bristol.

Pritchard, B., Curtis, A., Spriggs, J. and Le Heron, R. (Eds.) (2003) *Social Dimensions of the Triple Bottom Line in Rural Australia.* Australian Government Department of Agriculture, Fisheries and Forestry, Canberra.

Randall, E. (2002) Changing perceptions of risk in the European Union. In: *The Plan for a European Food Authority and the Politics of Risk in the European Union.* http://www.policylibrary.com/essays/RandallEFARisk/EFARisk1.htm accessed 13/8/2002.

Schusler, T.M., Decker, D.J. and Pfeffer, M.J. (2003) Social learning for collaborative natural resource management. *Soc. Nat. Resour.* **16**, 309–326.

Slovic, P. (1999) Trust, emotion, sex, politics and science: Surveying the risk-assessment battlefield. *Risk Anal.* **19**, 689–701.

Swan, K. (1987) *The Shire of Tallangatta: A History*. Tallangatta Shire Council, Tallangatta, Victoria.

Syme, G.J. and Eaton, E. (1989) Public involvement as a negotiation process. *J. Social Issues* **45**, 87–107.

Trettin, L. and Musham, C. (2000) Is trust a realistic goal of environmental risk communication? *Environ. Behav.* **32**, 410–427.

Index

Advances in Ecological Research
Volume 1–39

Cumulative List of Titles

Aerial heavy metal pollution and terrestrial ecosystems, **11**, 218

Age determination and growth of Baikal seals (*Phoca sibirica*), **31**, 449

Age-related decline in forest productivity: pattern and process, **27**, 213

Analysis and interpretation of long-term studies investigating responses to climate change, **35**, 111

Analysis of processes involved in the natural control of insects, **2**, 1

Ancient Lake Pennon and its endemic molluscan faun (Central Europe; Mio-Pliocene), **31**, 463

Ant-plant-homopteran interactions, **16**, 53

Anthropogenic impacts on litter decomposition and soil organic matter, **38**, 263

Arrival and departure dates, **35**, 1

The benthic invertebrates of Lake Khubsugul, Mongolia, **31**, 97

Biogeography and species diversity of diatoms in the northern basin of Lake Tanganyika, **31**, 115

Biological strategies of nutrient cycling in soil systems, **13**, 1

Bray-Curtis ordination: an effective strategy for analysis of multivariate ecological data, **14**, 1

Breeding dates and reproductive performance, **35**, 69

Can a general hypothesis explain population cycles of forest lepidoptera? **18**, 179

Carbon allocation in trees; a review of concepts for modeling, **25**, 60

Catchment properties and the transport of major elements to estuaries, **29**, 1

A century of evolution in *Spartina anglica*, **21**, 1

Changes in substrate composition and rate-regulating factors during decomposition, **38**, 101

The challenge of future research on climate change and avian biology, **35**, 237

Climate influences on avian population dynamics, **35**, 185

Climatic and geographic patterns in decomposition, **38**, 227

Climatic background to past and future floods in Australia, **39**, 13

The climatic response to greenhouse gases, **22**, 1

Coevolution of mycorrhizal symbionts and their hosts to metal-contaminated environment, **30**, 69

Communities of parasitoids associated with leafhoppers and planthoppers in Europe, **17**, 282

Community structure and interaction webs in shallow marine hardbottom communities: tests of an environmental stress model, **19**, 189

Complexity, evolution, and persistence in host-parasitoid experimental systems with *Callosobruchus* beetles as the host, **37**, 37